FEDER

Fish and Invertebrate Culture

FISH AND INVERTEBRATE CULTURE

Water Management in Closed Systems

Second Edition

STEPHEN SPOTTE

Vice President
Sea Research Foundation, Inc.
and
Director
Mystic Marinelife Aquarium
Mystic, Connecticut

A Wiley-Interscience Publication
John Wiley & Sons, New York • Chichester • Brisbane • Toronto

Library of Congress Cataloging in Publication Data

Spotte, Stephen
  Fish and invertebrate culture.

  "A Wiley-Interscience publication."
  Bibliography: p.
  Includes index.
  1. Aquarium water. I. Title. II. Title: Water
management in closed systems.

SF457.5.S66   1979   639'.3   78-10276
ISBN 0-471-02306-X

Printed in the United States of America

10 9 8 7 6 5 4 3 2 1

For my daughter Sara

Preface to the First Edition

In recent years there has been a worldwide upsurge of interest in maintaining freshwater and marine fishes and invertebrates in captivity. Aquarists have traditionally been interested in these animals from the standpoint of exhibition. But much of today's enthusiasm can be traced to the increased use of aquatic animals in research and also to the burgeoning science of aquaculture, in which marketable aquatic species are farmed on a commercial scale. In any case, successful results depend on sustaining and reproducing stable water conditions.

This book deals with the closed-system approach to culturing aquatic animals, in which the water is recycled and used again. In semiclosed and open systems, water is continuously discarded and replenished from a natural source. Mastery of closed systems has several advantages. It enables the culturist to raise marine organisms hundreds of miles from the sea, the advantage being that site selection for a saltwater laboratory, hatchery, or public aquarium need not be dictated by an available supply of natural seawater. Synthetic seawater can be used instead. Both freshwater and seawater closed systems can be operated in heavily polluted areas without depending on the local water supply. But most important, closed systems offer better environmental control. Natural water is subject to seasonal temperature fluctuations and may sometimes carry disease organisms, silt, pollutants, and undesirable animals that compete with the cultured specimens for space and nutrients or even prey upon them.

Aquatic animals change the chemistry of their culture water. Their physiology is in turn affected by these changes, usually in adverse ways. The purpose of this book is to show how to culture fishes and invertebrates in closed systems by controlling the chemical factors in the environment bringing about changes in normal physiology. The major concept presented is that accumulating toxic metabolites are the primary limiting factors in aquatic animal culture and that removing them on a continuous basis should be the foremost objective in routine water management.

Previous books dealing with culture problems of aquatic animals have emphasized a biological viewpoint. Osmotic imbalance, thermal shock, partial asphyxiation, lowered disease resistance, stunting, and loss of fecundity are typical problems of captivity, but they can seldom be corrected by treating the biological effects. These and many other maladies are often the tangible results of deteriorating water quality.

This book has been designed as a water management handbook for the researcher, teacher, or advanced student maintaining a modest collection of animals in the laboratory. It is also written for the fishery biologist and hatchery manager responsible for thousands of animals in various stages of development. The size of a culture system has little bearing on the isolation and solution of water management problems. These are essentially the same in a 10-gallon aquarium or a 10-million-gallon hatchery.

The term "aquatic animals," as used throughout the book, refers collectively to fishes and invertebrates of both freshwater and marine origins. In general, the same water quality standards and water management techniques apply to both groups of animals and to both types of environment.

The book is divided into two sections. Part I concentrates on the unfavorable effects that animals have on captive water. Part II deals with the effects of the culture water on the physiology of the animals. Both parts stress theoretical principles as well as practical water management techniques. It is assumed that the reader will have a stronger background and orientation in biology than in chemistry. Therefore, only theoretical data relating to the practical aspect of the subject are presented. Rigorous thermodynamic and mathematical functions of important reactions are avoided because they serve no practical purpose. Laboratory tests in the last chapter have been simplified for the same reason. Finally, the literature cited is meant to be representative of published work, rather than a complete survey.

STEPHEN SPOTTE

Niagara Falls, New York
May 1970

Preface to the Second Edition

One editor who reviewed the first edition of this book noted that the title *Fish and Invertebrate Culture* was a misnomer because the text contained no information about how to rear aquatic animals. He was right, and I apologize to any readers who were misled. Unfortunately, succeeding editions of a book do not easily lend themselves to changes in title, but I shall state here that no methods of rearing fishes and invertebrates are given in this edition either. The subject is the same as before: water quality control in closed-system environments. This time I have limited the field to aquariums, which ordinarily have low densities of animals. Hatchery and aquaculture installations often have different problems because animal densities are higher.

The book has been rewritten almost entirely, but every effort has been made to keep the text simple, brief, and practical. Theoretical information is presented only if pragmatism is served. As in the first edition, literature cited is meant to be representative, and not a complete survey.

STEPHEN SPOTTE

Mystic, Connecticut
February 1979

Acknowledgments

I would like to acknowledge the assistance of four dedicated people who helped in the preparation of the second edition of this book. Gary Adams of Thames Valley State Technical College, Norwich, Connecticut, helped immeasurably with the calculations and other portions of the manuscript that dealt with mathematics and chemistry. James W. Atz of the American Museum of Natural History, New York City, Carol E. Bower, Institute for Aquarium Studies, Hartford, Connecticut, and William E. Kelly, Aquarium Systems, Inc., Eastlake, Ohio reviewed the text and offered many valuable criticisms and suggestions. Lee C. Eagleton of The Pennsylvania State University, University Park, Pennsylvania, prepared much of the section on airlift pumping. Paul Gaj of Mystic Marinelife Aquarium, Mystic, Connecticut, redrew many of the figures in the first edition, adapted newer illustrations from the published work of other investigators, and converted my crude sketches of the original figures for this edition into art.

S. S.

Credits

Figure 1-3: A. Kawai, Kyoto University, Kyoto, Japan. *Figure 1-5*: Y. Yoshida, Misaki Marine Biological Institute, Kyoto University, Mazizuru, Japan. *Figure 1-6*: *Petfish Monthly*, London, England. *Figure 2-2*: reprinted with permission of *Environmental Science and Technology* **5**: 1105–1112, 1971, copyright by the American Chemical Society. *Figures 2-3, 3-13, 3-14*: reprinted with permission of the American Water Works Association, Inc. *Figure 3-3*: Universität Karlsruhe, West Germany. *Figures 3-8, 4-6, 4-4*: copyright 1967, 1973, 1977 by the Water Pollution Control Federation. *Figure 3-9*: reprinted with permission of *Water and Sewage Works* magazine. *Figures 3-10, 3-11*: E. Sander, Am Osterberg, West Germany. *Figures 3-12, 3-15*: reprinted with permission of *Aquaculture*, Elsevier Scientific Publishing Co., Amsterdam, Holland. *Figures 4-2, 7-1*: reprinted with permission of John Wiley and Sons, New York. *Table 4-1*: reprinted with permission of the American Fisheries Society. *Figure 5-5*: reprinted with permission of the Wistar Institute of Anatomy and Biology, Philadelphia, Pennsylvania. *Figure 5-6*: reprinted with permission of *Biological Bulletin*, Marine Biological Laboratory, Woods Hole, Massachusetts. *Figures 4-3, 7-2*: reprinted with permission of Pergamon Press Ltd., Oxford, England. *Table 6-1*: reprinted with permission of Academic Press, New York. *Figure 6-1*: William A. Anikouchine, Richard W. Sternberg, *The World Ocean: An Introduction to Oceanography*, copyright 1973, p. 75, by permission of Prentice-Hall, Englewood Cliffs, New Jersey. *Table 7-1*: reprinted with permission of *Journal of Chemical Engineering Data* **8**: 464–468, 1963, copyright by the American Chemical Society. *Table 9-1*: reprinted with permission of *Journal of the Fisheries Research Board of Canada*.

Contents

Fish and Invertebrate Culture

CHAPTER 1

Biological Filtration

Biological filtration includes some of the most important biochemical processes that take place in closed-system aquariums. As defined and used here, *biological filtration* is the mineralization, nitrification, and dissimilation of nitrogenous compounds by bacteria suspended in the water and attached to gravel and detritus in the filter bed. Organisms adapted to carry out these functions are always present as part of the normal filter bed biota. Mineralization and nitrification alter the chemical state of nitrogenous compounds in the water but do not remove any nitrogen from solution. This can be accomplished biologically only by dissimilation (Section 1.3).

Biological filtration is the first of four separate processes used in aquarium water management schemes. The other three are mechanical filtration, physical adsorption, and disinfection, which are the subjects of the following three chapters. The whole scheme is illustrated diagrammatically in Fig. 1-1. The aquarium nitrogen cycle, depicting mineralization, nitrification, and dissimilation is shown in Fig. 1-2.

1.1 MINERALIZATION

Heterotrophic and autotrophic bacteria are the major groups present in aquariums. Heterotrophic species utilize nitrogenous organic compounds excreted by the animals as energy sources and convert them to simple

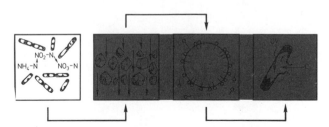

Figure 1-1. Position of biological filtration in a water management scheme. The four processes, from left to right, are biological filtration, mechanical filtration, physical adsorption, and disinfection. Original.

2

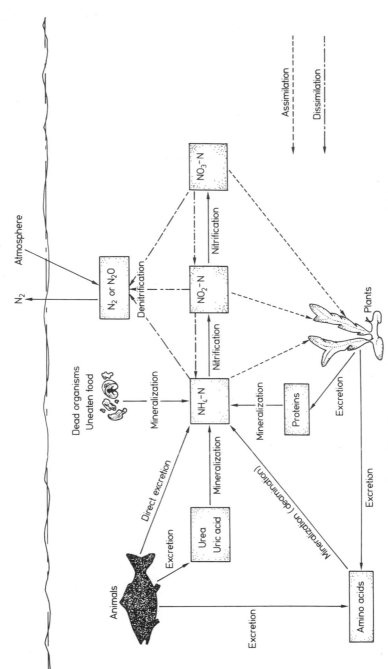

Figure 1-2. The nitrogen cycle in closed-system aquariums. Original.

compounds, such as ammonia.* The mineralization of these organics is the first stage in biological filtration. In the case of nitrogenous organic compounds, mineralization may begin by the decomposition of proteins and nucleic acids to produce amino acids and nitrogenous organic bases. Deamination is a mineralization process in which an amino group is split off to form ammonia. An example of deamination is the breakdown of urea to produce free ammonia (NH_3), as shown in eq. 1:

$$O=C \begin{array}{c} \diagup NH_2 \\ \diagdown NH_2 \end{array} + H_2O \rightarrow CO_2 + 2NH_3 \tag{1}$$

The same reaction can also proceed by a purely chemical process, but the deamination of amino acids and related compounds requires the presence of bacteria.

1.2 NITRIFICATION

Once organics have been converted to an inorganic state by heterotrophs, biological filtration shifts to the second stage which is *nitrification*. This process is defined as the biological oxidation of ammonia to nitrite (NO_2^- measured as NO_2-N) and nitrate (NO_3^- measured as NO_3-N). Nitrification is carried out mainly by autotrophic bacteria. Autotrophic organisms, unlike heterotrophs, can utilize inorganic carbon (mainly CO_2) as a source of cellular carbon. *Nitrosomonas* and *Nitrobacter* are probably the principal genera of autotrophic nitrifying bacteria in freshwater, brackish water, and seawater aquariums. *Nitrosomonas* oxidizes ammonia to nitrite (eq. 2); *Nitrobacter* oxidizes nitrite to nitrate (eq. 3).

$$NH_4^+ + OH^- + 1.5O_2 \rightarrow H^+ + NO_2^- + 2H_2O \tag{2}$$

$$\Delta G° = -59.4_{kcal}$$

$$NO_2^- + 0.5O_2 \rightarrow NO_3^- \tag{3}$$

$$\Delta G° = -18.0_{kcal}$$

Both reactions show a fall in free energy. The significance of eqs. 2 and 3 is the conversion of toxic ammonia to nitrate, which is far less toxic. The efficiency of the nitrification process is affected by six factors: (*1*) the

* The term "ammonia" refers to the sum of ammonium ion (NH_4^+) and free ammonia (NH_3), determined analytically as total NH_4-N.

presence of toxic compounds in the water; (2) temperature; (3) pH; (4) the concentration of dissolved oxygen; (5) salinity; and (6) surface area of the filtrant material.

Toxic Compounds

Many chemical compounds inhibit nitrification under certain conditions. When added to aquarium water, they affect the filter bed bacteria in one of two ways: either (1) the growth and proliferation of these organisms are repressed; or (2) there is no effect on growth and proliferation, but the metabolism of the cells is interrupted, which prevents them from reaching full oxidizing capacity.

Collins et al. (1975, 1976) and Levine and Meade (1976) reported that many commonly used antibacterial agents and parasiticides used in treating fish diseases had no effect on nitrification in freshwater aquariums; others were toxic to various degrees. No parallel studies have been made using seawater, and the experimental results of these authors should not be extrapolated for application to saline environments.

The data of the three reports cited above are summarized in Table 1-1. The results are not comparable in some respects because of differences in technique. Collins and his co-workers measured the effects of the compounds they tested by taking water samples directly from aquarium tanks that contained fish and viable filter beds. Levine and Meade used inoculates of mixed bacteria and their work was done in culture. Levine and Meade acknowledged that their results perhaps showed more sensitivity than might be the case under actual operating conditions. For example, their tests showed formalin, malachite green, and nifurpirinol (Furanace®) to be mildly toxic to nitrifying bacteria, whereas Collins et al. (1975, 1976) showed the same compounds to be harmless. Levine and Meade suggested that the difference was attributable to the greater percentage of autotrophs in their inoculates, and that perhaps the low inhibition thresholds for these substances did not exist in the presence of greater numbers of heterotrophs and higher concentrations of dissolved organic matter.

From the data in the table, there is little doubt about the inhibiting properties of erythromycin, chlorotetracycline, methylene blue, and sulfanilamide in freshwater. Methylene blue appeared to be the most toxic of the substances tested. Mixed results were obtained with chloramphenicol and potassium permanganate.

Both Collins et al. (1975) and Levine and Meade (1976) found that copper did not inhibit nitrification significantly (Table 1-1). This was perhaps the result of free copper ions having been chelated with dissolved organic matter in solution. Tomlinson et al. (1966) determined that the effects of the heavy metals Cr, Cu, and Hg on *Nitrosomonas* in pure culture

Table 1-1. Effects of Commonly Used Antibacterial Agents and Parasiticides on Nitrification in Freshwater Aquariums at Therapeutic Levels. From data in Collins et al. (1975, 1976) and Levine and Meade (1976)

Compound	Concentration (mg l^{-1})	% Inhibition	Source‡
Chloramphenicol	50	0	b
	50	84	c
Oxytetracycline	50	0	b
Sulfamerazine	50	0	b
Sulfanilamide	25	65	c
Erythromycin	50	100	b
Nifurpirinol	1	0	b
	0.1	20	c
	4	44	c
Chlorotetracycline	10	76	c
Formalin	25*	0	a
	15	27	c
Malachite green	0.1	0	a
Formalin + malachite green	25 + 0.1	0	a
Methylene blue	5	100	a
	1	92	c
Copper sulfate	1†	0	a
	5	0	c
Potassium permanganate	4	0	a
	1	86	c

* Equivalent to 10 mg l^{-1} formaldehyde.
† Hardness = 30 mg CaCO$_3$ l^{-1}.
‡ a, Collins et al. (1975); b, Collins et al. (1976); c, Levine and Meade (1976).

were far more severe than the effects in activated sludge. They postulated that such results were probably due to the formation of chemical complexes between the metals and organics. Long-term effects of heavy metals were more severe than short-term effects. Possibly because adsorption sites on the organic molecules were used up.

Temperature

Many species of bacteria can survive large changes in temperature, although their activities may be affected temporarily. A period of adjustment, called a *time lag*, is often evident if the temperature has been altered abruptly. Time lags are usually seen when the temperature is lowered suddenly; increases in temperature ordinarily speed up biochemical activities and therefore may not produce a time lag. Srna and Baggaley (1975) studied the kinetics of nitrification in seawater aquariums. A rise of 4°C increased ammonia and nitrite oxidation by 50% and 12%, respectively, compared with calculated values. Lowering the temperature 1°C slowed down the oxidation rate of ammonia by 30%, and a 1.5°C decrease reduced the rate of nitrite oxidation by 8%, compared with calculated values.

pH

Kawai et al. (1965) found that lowering the pH from 9.0 was more detrimental to nitrification in seawater than in freshwater aquariums. They attributed this to the normally lower pH values of freshwater. Saeki (1958) found that the oxidation of ammonia in freshwater aquariums was inhibited by low pH. The ideal pH was 7.8 for ammonia oxidation and 7.1 for conversion of nitrite. Saeki considered the best range for freshwater culturing to be 7.1–7.8, insofar as nitrifying bacteria were concerned. Srna and Baggaley (1975) determined that the marine nitrifiers used in their study functioned best at pH 7.45, with an effective range of 7.0–8.2.

Dissolved Oxygen

A filter bed can be compared with a huge, respiring organism. When functioning properly, it consumes a considerable amount of oxygen. The oxygen consumption of microorganisms in a water system is measured in terms of BOD (biological oxygen demand). The BOD of a filter bed is partly a function of nitrification, but most is caused by the activities of heterotrophic bacteria. Hirayama (1965) showed that if the BOD of a filter bed was high, a sizeable population of nitrifiers was at work. Hirayama filtered aquarium seawater through a column of sand from an old filter bed. Before entering the column, the water had a dissolved oxygen level of 6.48 mg l^{-1}; after passing through 48 cm of sand, it measured 5.26 mg l^{-1}. At the same time, ammonia decreased from 238 to 140 meq l^{-1}, and nitrite from 183 to 112 meq l^{-1}.

Both aerobic and anaerobic bacteria are found in filter beds, but aerobic forms predominate in well-aerated aquariums. The growth and activities of anaerobic bacteria are inhibited in the presence of oxygen, and adequate circulation through a filter bed keeps them in check. When the oxygen tension of the aquarium decreases, anaerobes proliferate, or convert over from aerobic respiration. Many metabolites produced during anaerobic activity are toxic. Mineralization can take place when oxygen tensions are low, but the mechanisms and end products are different. In anaerobic situations, the mechanisms are fermentative, rather than oxidative. During mineralization, this results in the formation of organic acids instead of bases, along with carbon dioxide and ammonia. These substances, plus hydrogen sulfide, methane, and several others are what give a suffocating filter bed its familiar putrid odor.

Salinity

Many species of bacteria can live in waters that vary considerably in ionic strength, provided they are acclimated to the different conditions gradually. ZoBell and Michener (1938) found that most of the bacteria isolated from seawater in their laboratory could also be grown in freshwater. Many could even survive the change directly. Of 12 species that appeared superficially to be "marine" bacteria, all were converted successfully to freshwater by gradual dilution of their seawater medium in increments of 5%.

Filter bed bacteria are fairly tolerant of salinity changes, although repression of activity is measurable if the change is large and abrupt. Srna and Baggaley (1975) demonstrated that a decrease in salinity of 8‰ (parts per thousand) and an increase of 5‰ failed to affect the rate of nitrification in seawater aquariums. Kawai et al. (1965) found that nitrifying activities in a seawater aquarium were greatest when the water was at normal salinity. Nitrification diminished as the solution was either diluted or concentrated, although some activity remained even after the salinity was doubled. Nitrification in a freshwater aquarium was greatest before the addition of sodium chloride. Activity stopped altogether when the salinity was raised to a level equivalent to full-strength seawater.

There is some evidence that salinity affects nitrification rates and even the quantities of end products at equilibrium . Kuhl and Mann (1962) showed that nitrification proceeded more rapidly in freshwater than in seawater aquariums, with greater quantities of nitrite and nitrate being formed in seawater. Kawai et al. (1964) obtained similar results, which are summarized in Fig. 1-3.

Surface Area

Kawai et al. (1964) determined that the nitrifiers in a filter bed were 100 times more plentiful than those suspended in the water. This indicated that an important factor in nitrification was the surface area that is available for attachment by bacteria. The greatest surface area in an aquarium is provided by the gravel grains, and most of the nitrification takes place in the upper gravel layer (Kawai et al. 1965, Yoshida 1967), as shown in Fig. 1-4. Kawai et al. (1965) found that there were 10^5 ammonia oxidizers per gram of sand in the top portions of the filter bed in a seawater aquarium,

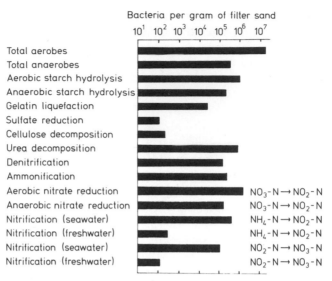

Figure 1-3. Populations of filter bed bacteria in small freshwater and seawater aquariums after 134 days. Redrawn from Kawai et al. (1964).

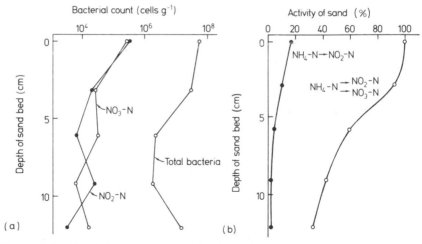

Figure 1-4. Populations (*a*) and activities (*b*) of nitrifying bacteria at different depths in the filter bed of a seawater aquarium. Redrawn from Yoshida (1967).

and that the number of nitrite oxidizers was 10^6. The population of each type fell by 90% at a depth of only 5 cm.

The size and shape of the gravel is also important. Small grains have more surface area for attachment by bacteria than comparable masses of large grains, although very fine gravels are undesirable if they restrict water circulation. The size and surface area relationship is easily demonstrated. Six cubes, each with a mass of 1.0 g, have a total of 36 unit surfaces, whereas one cube with a mass of 6.0 g has only 6 surfaces, each larger than the individual surfaces of the small cubes. The total surface area of the six l-g cubes is 3.3 times greater than the surface area of the single 6-g cube. Saeki (1958) suggested that gravel ranging in size from 2–5 mm was best for use in aquarium filter beds.

Angular gravel has more surfaces than round types. A sphere has the smallest surface area per volume of any geometric shape. Coarse, angular gravels are preferable to smooth, water-worn varieties.

Accumulated detritus provides additional surfaces and improves nitrification. According to Saeki (1958), 25% of all nitrification in aquariums is accomplished by bacteria attached to detritus.

1.3   DISSIMILATION

Inorganic nitrogen is oxidized to higher oxidation states as the result of nitrification. *Dissimilation*, or "nitrate respiration," works in the opposite way by reducing the end products of nitrification back to lower states of oxidation. In terms of total activity, the oxidation of inorganic nitrogen far exceeds its reduction, and nitrate accumulates. Other than dissimilation, which allows some nitrogen to be driven into the atmosphere, excess inorganic nitrogen can be removed from solution only by administering regular partial water changes, culturing plants, or the use of ion exchange resins. This last technique is effective only in freshwater, as is explained in Section 3.3.

Dissimilation is mainly an anaerobic process that occurs in those portions of an aquarium filter bed that are deficient in oxygen. Bacteria that carry out dissimilatory activities either are true anaerobes, or aerobes that can switch to anaerobic respiration under anoxic conditions. Ordinarily, these organisms are heterotrophic; several species of *Pseudomonas*, for example, can reduce nitrate ion (NO_3^-) in low-oxygen environments (Painter 1970).

During anaerobic respiration, dissimilatory bacteria utilize NO_3^- in place of oxygen. In doing so, they reduce nitrate to other forms with lower oxidation numbers: nitrite, ammonia, nitrous oxide (N_2O), or molecular nitrogen (N_2). The particular end product depends on the species of bacterium involved. When inorganic nitrogen is reduced completely—that is, to N_2O or N_2—the dissimilation process is known as *denitrification*. In the completely reduced state, some nitrogen can be removed from the water and released into the atmosphere when its partial pressure in solution exceeds its partial pressure in the air above the aquarium. Denitrification thus lowers the level of inorganic nitrogen in aquariums, whereas mineralization and nitrification do not.

1.4 THE "CONDITIONED" AQUARIUM

A *conditioned* aquarium is one in which the filter bed bacteria are in equilibrium with the routine input of their energy sources. Nitrification can be used to determine when a new aquarium becomes conditioned and suitable for maintaining delicate aquatic organisms. At first a high ammonia level is the limiting factor. Ammonia ordinarily subsides within 2 weeks under warm-water conditions (ambient temperature about 15°C or higher), and after a considerably longer period in cold water (ambient temperature below about 15°C). An aquarium may be ready to receive animals within the first 2 weeks, but it is not completely conditioned because many of the important groups of bacteria have not yet stabilized. Kawai et al. (1964) described the bacterial population of a warm seawater aquarium after 3 months as follows (the results are depicted graphically in Fig. 1-3):

1. *Total aerobic.*   Increased tenfold within 2 weeks after addition of fishes. Maximum population density of 10^8 organisms per gram of filter sand was reached after 2 weeks. The population stabilized at 10^7 g^{-1} of filtrant after 3 months.
2. *Protein decomposing.*   Original population started at 10^3 g^{-1} of filter sand and increased 100-fold after 4 weeks. Stabilization at 10^4 occurred after 3 months. The reason for the dramatic increase was thought to be caused by the food (fish flesh), which was high in protein.
3. *Starch decomposing.*   Original population was 10% of the total number of bacteria. There was a gradual increase, then a decline after 4 weeks. The population stabilized at 1% of the total population after 3 months.
4. *Nitrifiers.*   Maximum density of nitrite formers was reached after 4 weeks and that of nitrate formers after 8 weeks. Nitrite formers were present in greater numbers than nitrate formers after 2 weeks. Stabilization occurred at 10^5 and 10^6 organisms per gram of filtrant, respectively.

There is a time lag between the fall of ammonia and the oxidation of nitrite in the initial stages of nitrification. This is because the growth of *Nitrobacter* is inhibited by the presence of ammonia (Anthonisen et al. 1976, Lees 1952). Efficient oxidation of nitrite does not take place until most of the ammonia has been converted by *Nitrosomonas,* as shown in Fig. 1-4. Similarly, nitrite must peak before nitrate starts to accumulate.

The high ammonia level in a new aquarium may result from a population imbalance between heterotrophic and autotrophic bacteria. When a new aquarium is put into operation, the growth of heterotrophic species at first exceeds the growth of autotrophs. Much of the ammonia that is produced from mineralization is used by part of the heterotrophic population. In other words, there is no sharp dividing line between heterotrophic and autotrophic ammonia utilization. Significant ammonia oxidation by nitrifiers proceeds only after the heterotrophic population has subsided and stabilized (Quastel and Scholefield 1951).

The numbers of bacteria in a new aquarium are a significant factor only until each type has stabilized. Subsequently, increases in the metabolic activities of the individual cells compensate for fluctuations in energy sources, but there is no additional increase in the quantity of cells. Work by Quastel and Scholefield (1951) and Srna and Baggaley (1975) showed that the population densities of nitrifying bacteria occupying a given number of surfaces are relatively constant and independent of the concentration of the available energy source.

The total oxidizing capacity of the bacterial population in a conditioned aquarium is geared to a stable daily input of oxidizable substrates. Sudden increases in the number of animals, their masses, and the quantity of food added each day often produce measurable increases in the ammonia and nitrite levels. These may persist until the bacteria equilibrate with the new conditions.

The extent of increased levels of ammonia and nitrite depends on how much the additional nutrient load has stressed the carrying capacity of the water system (Section 1.5). If the increase in the animal and food load is still below the maximum carrying capacity, equilibrium with the new conditions is usually attained within 3 days in warm water, and after a significantly longer time in cold water. If the additional load pushes the aquarium beyond its maximum carrying capacity, the results are permanently increased ammonia and nitrite levels.

Mineralization, nitrification, and dissimilation are processes that follow one another more or less in sequence in a new aquarium. Once conditions reach equilibrium, they all occur almost simultaneously. In a conditioned aquarium, the measurable ammonia (as total $NH_4\text{-}N$ is less than 0.1 mg

l^{-1}, and any measurable nitrite is often the result of dissimilation. These levels remain constant and there are no time lags in the conversion process because all energy sources are utilized rapidly.

1.5   DESIGN

The design of subgravel filters can be divided into three sections: (*1*) general design criteria; (*2*) a discussion of how to construct filter plates; and (*3*) criteria for designing airlift pumps. Airlift pump design is discussed in Section 5.1 because it falls under the category of gas exchange.

General Design Criteria

A subgravel filter provides the only biological and mechanical filtration needed in most aquarium tanks, even very large ones. Design criteria for biological and mechanical filtration are therefore the same: (*1*) surface area of the filter bed should equal the surface area of the aquarium tank; (*2*) grain size of the gravel should be in the range of 2–5 mm; (*3*) gravel must be carefully graded to assure uniform grain size; (*4*) the filter bed should not vary in depth; (*5*) shape of the gravel grains should be rough and angular; (*6*) flow rate through the filter bed should be about 0.7×10^{-3} m sec^{-1} (1.0 gal ft^{-2} min^{-1}); and (*7*) minimum depth of the filter bed should be 7.6 cm. Number *1* requires further discussion. All are simply rules of thumb that have proved to be reliable.

The distribution of bacteria in a filter bed is a direct function of depth. The efficiency of nutrient conversion is affected by depth only indirectly. Kawai et al. (1965) determined that heterotrophic bacteria in a seawater aquarium were most numerous at the surface of filter beds (about 10^8 g^{-1} of filter sand), and decreased about 90% 10 cm into the bed. The same trend held true for autotrophic species, as mentioned before. The surface sand populations of ammonia and nitrite oxidizers, which were 10^5 and 10^6, respectively, fell by 90% at a depth of only 5 cm. Based on these findings, Kawai et al. (1965) recommended that filter beds be designed with large surface areas and shallow depths. Yoshida (1967) reported that maximum activity by nitrifying bacteria in seawater aquariums occurred in the upper layer (Fig. 1-4). Activity diminished rapidly with increasing depth of the bed. A design in which the surface area of the filter bed equals that of the aquarium tank is therefore a basic requirement.

Hirayama (1965) demonstrated that the effect of filter bed depth on

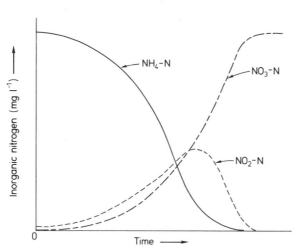

Figure 1-5. Stylized nitrification curves showing the sequential rise and fall of ammonia and nitrite and the rise of nitrate in closed-system aquariums. Redrawn from Anthonisen et al. (1976).

nutrient conversion in aquarium water was indirect when OCF (oxygen consumed during filtration) was used as an index. OCF can be interpreted to mean BOD min^{-1}. Conversely, the time taken for the water to pass through the bed could be directly correlated with OCF. Hirayama's data showed that the apparent effects of depth are misleading because the time required for water to pass through a filter bed is proportional to the depth. To prove the point, Hirayama designed an experiment in which culture water was passed through four filters that differed only in depth. The time required for water to move through the filtrant was made constant by adjusting the flow rate. In the end, it was shown that OCF values were the same, even though the depths of the filters differed. Thick beds thus need faster flow rates than thin ones.

Filter Plates

Filter plates must form a false bottom and suspend the gravel above the floor of the aquarium tank. It is important to seal the edges of the plate to the walls of the tank to keep gravel from working underneath. Filter plates for large aquariums can be made of any material that is sturdy, porous, and inert in water. Fiberglass-reinforced plastic roofing panels and fiberglass-reinforced epoxy industrial grating are two materials that are

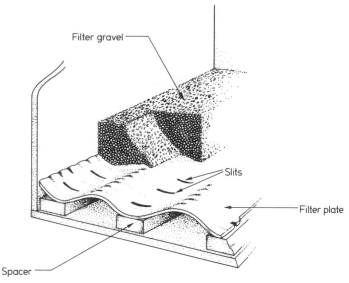

Filter gravel

Slits

Filter plate

Spacer

Figure 1-6. Cutaway section of an aquarium tank showing a filter plate made of fiberglass-reinforced plastic roofing material. Redrawn from Anonymous (1971).

used routinely at Aquarium of Niagara Falls and Mystic Marinelife Aquarium. Roofing panels are available at lumberyards and hardware stores. Industrial grating is manufactured by Joseph T. Ryerson and Son, Inc.*

When roofing panels are used, slits should be cut into them crosswise at right angles to the ribs using a table saw equipped with a blade for cutting plastic. The slits should be about 1 mm wide, 2.5 cm long, and 5 cm apart, as shown in Fig. 1-6. Afterward, panels are placed *slits down* in the aquarium tank and sealed where the edges meet the walls. The best sealant is fiberglass tape (5 cm wide) embedded in clear silicone, such as Dow Corning RTV 732† or equivalent. After the silicone hardens, gravel can be added to the desired depth and spread evenly over the filter plate.

When using grating, saw the material to the required dimensions, then lay plastic flyscreen on top. Tie the screen in place with monofilament fishing line or stainless steel wire. Afterward, seal the edges of the grate with silicone where it meets the walls of the aquarium tank. Both roofing and grating must be supported above the bottom of the aquarium tank

---

* Joseph T. Ryerson and Son, Inc., Box 484, Jersey City, N.J. 07303
† Dow Chemical U.S.A., Midland, Mich. 48640

with spacers. Any material can be used so long as it is inert and sturdy, including concrete bricks or semicircles of PVC pipe cut to the right length and stood on edge. The important thing is that water must be able to circulate freely around the spacers. Concrete bricks should be sprayed with a prime coat of epoxy paint followed by two finish coats, particularly in seawater applications. This keeps the concrete in the bricks from eroding. The spacers need not be attached to the bottom of the aquarium tank or to the filter plate.

Carrying Capacity

An important aspect of biological filtration is *carrying capacity,* which is defined as the animal load that an aquarium can support. Hirayama (1966a) derived the following formula for calculating carrying capacity in seawater aquariums:*

$$\sum_{i=0}^{p} \frac{10W_i}{\dfrac{0.70}{V_i}} + \frac{0.95 \times 10^3}{G_i D_i} \geqq \sum_{j=1}^{q} (B_j{}^{0.544} \times 10^{-2}) + 0.051F \qquad (4)$$

Any results should be used as guidelines only. The left-hand expression represents the oxidizing capacity of the filter bed (OCF) measured as milligrams of O_2 consumed per minute, where W = the surface area of the filter bed (m²), V = filtering rate, or flow rate of water moving through the filter bed (cm min⁻¹), D = gravel depth (cm), and p = the number of filters serving the aquarium. In the above equation (actually an inequality), G represents the grain size coefficient of the gravel grains. This is determined by

$$\frac{1}{R_1}x_1 + \frac{1}{R_2}x_2 + \frac{1}{R_3}x_3 + \ldots + \frac{1}{R_n}x_n \qquad (5)$$

where R = the mean grain size of each fraction of gravel in the filter bed (if the gravel is graded) in millimetres, and x = the percentage weight of each fraction.

The right-hand expression of the inequality (eq. 4) represents the rate of "pollution" by the animals. It is also expressed in mg O_2 min⁻¹. In this expression, B = the body masses of the individual animals (g), F = the

---

* The same relationship may also hold true for freshwater aquariums, but this needs verification.

amount of food (g) entering the aquarium daily, and q = the number of animals being maintained.

As seen from the formula, the oxidizing capacity of the filter bed must be greater than, or equal to, the rate of "pollution" by the animals. It is also important to note that *as the masses of the individual animals decrease, the carrying capacity of the aquarium decreases.* In other words, carrying capacity is not simply a function of the total animal mass. An aquarium that can support a single 100-g fish cannot necessarily support 10 fishes each with a mass of 10 g. Assume, for example, that in a hypothetical aquarium $W = 0.35$ m^2, $V = 10.5$ cm min^{-1}, and $D = 36$ cm. If the gravel is all the same grade and $R = 4$ mm, then from eq. 5, $G = \frac{1}{4} \times 100 = 25$. Substitution of these values into the left-hand expression of the original inequality gives the OCF value, which is equivalent to BOD min^{-1}.

$$\frac{10(0.36)}{\dfrac{0.70}{10.5} + \dfrac{0.95 \times 10^3}{25(36)}} = \frac{3.6}{0.067 \times \dfrac{950}{900}}$$

$$= \frac{3.6}{0.067 + 1.055} = \frac{3.6}{1.122} = 3.2 \text{ mg OCF min}^{-1}$$

Assume further that fishes of 200-g each are being maintained and that they are fed at 5% of their individual body masses per day. From the right-hand expression of eq. 4, X represents OCF; therefore,

$$X = \sum_{j=1}^{q} (B_j^{0.544} \times 10^{-2}) + 0.051F \qquad (6)$$

Table 1-2 shows the value of X for one fish as a function of mass in grams and feeding rate as a percentage of body mass per day. From the table, feeding a 200-g fish 5% of its body mass daily corresponds to a "pollution load" of 0.69 mg OCF min^{-1}. The value for $q = X/0.69 = 3.2/0.69 = 4.6$ fishes, indicating that four fishes can be maintained in the aquarium. Care must be taken to be conservative when using this procedure. "Pollution load" changes as animals grow, and the carrying capacity of the filter bed may be exceeded suddenly when a fish dies or conditions should become anoxic.

As a second example, determine if the same aquarium could support 10 50-g fishes and one that weighs 600 g, all of which are fed at 5% of their individual body masses daily. As seen in Table 1-2, the "pollution load" is $10(0.21) + 1(1.85) = 3.95$ mg OCF min^{-1}. The answer is no, because the "pollution load" exceeds the carrying capacity, which is 3.2 mg OCF min^{-1}.

Table 1-2. "Pollution Load" as a Function of the Mass of a Fish and Its Feeding Rate. Calculated from the Right-hand Expression of Eq. 4

Body Mass (g)	Feeding Rate (% of Body Mass da^{-1})				
	0.0%	2.5%	5.0%	7.5%	10.0%
30	0.06	0.10	0.14	0.18	0.22
40	0.07	0.13	0.18	0.23	0.28
50	0.08	0.15	0.21	0.28	0.34
60	0.09	0.17	0.25	0.32	0.40
80	0.11	0.21	0.31	0.41	0.52
100	0.12	0.25	0.38	0.50	0.63
150	0.15	0.34	0.54	0.73	0.92
200	0.18	0.43	0.69	0.94	1.20
250	0.20	0.52	0.84	1.16	1.48
300	0.22	0.61	0.99	1.37	1.75
400	0.26	0.77	1.28	1.79	2.30
500	0.29	0.93	1.57	2.21	2.84
600	0.32	1.09	1.85	2.62	3.38
800	0.38	1.40	2.42	3.44	4.46
1000	0.43	1.70	2.97	4.25	5.53
1500	0.53	2.45	4.36	6.27	8.18
2000	0.62	3.17	5.72	8.27	10.80
3000	0.78	4.60	8.43	12.30	16.10
4000	0.91	6.01	11.10	16.20	21.30
5000	1.03	7.40	13.80	20.20	26.50
6000	1.14	8.79	16.40	24.10	31.70
8000	1.33	11.50	21.70	31.90	42.10
10,000	1.50	14.20	27.00	39.70	52.50
20,000	2.19	27.70	53.20	78.70	104.10
30,000	2.72	40.90	79.20	117.50	155.70
40,000	3.19	54.20	105.20	156.19	207.20

1.6 MANAGEMENT PRACTICES

The proper management of a biological filter involves: (*1*) conditioning it quickly and effectively when it is new; and (*2*) giving it the right care once it has become conditioned.

New Aquariums

It is best to *overcompensate* a new filter bed when conditioning it; that is, to condition it with a slightly greater animal load than it will ultimately support. Overcompensation eliminates later increases in ammonia and nitrite when more animals are added.

Only hardy animals should be used to condition a new aquarium. Animals that are highly susceptible to ammonia poisoning should not be added until nitrification is fully established. Turtles are excellent to start out new aquariums. They are less affected by ammonia than fishes and many invertebrates, yet they supply the organic material needed to initiate mineralization and nitrification. Marine turtles and diamondback terrapins are suitable for conditioning brackish water or seawater aquariums, and any of the common freshwater species, such as sliders, snappers, and map turtles are ideal for freshwater aquariums. Among the fishes, moray eels, groupers, carp, and many of the catfishes are notably tolerant of ammonia. Lobsters and crabs are invertebrates that can tolerate high ammonia concentrations without suffering ill effects.

Adding specimens a few at a time is always a good technique. If no hardy animals are available, and if the species to be cultured are sensitive to ammonia poisoning, the animal load can be built up gradually to maximum density. For example, if it is necessary to keep the ammonia level below 0.2 mg total NH_4-N l^{-1} at all times, the animal population can be increased slowly at a rate that will not cause the ammonia to exceed this level. The technique is to monitor the ammonia level continuously and not add specimens faster than the population of nitrifiers can stabilize the ammonia at 0.2 mg l^{-1} or less. This technique involves much laboratory work and prolongs conditioning time. The first method—overcompensating with hardy animals—is quicker and more practical.

The time lags in cold water are longer because the growth of bacteria is slower at low temperatures. The biological filtration processes can be speeded up by maintaining the aquarium at warm-water temperatures with warm-water animals until nitrification is established. The animals can then be removed, the temperature dropped, and a similar (preferably lower) weight of cold-water specimens added. Sometimes on adding the cold-water animals, slight increases in ammonia and nitrite are observed, even if the aquarium has been overcompensated initially. These elevated concentrations ordinarily subside after several days, if the temperature disparity is not too great, indicating that the bacteria have adapted to the cold. Such increases can be minimized by allowing at least 96 hr for the filter bed bacteria to adjust to the lower temperature before adding the cold-water animals.

One reliable method of accelerating the conditioning process is to inoculate the new aquarium with an established population of filter bed bacteria. A portion of the surface gravel and detritus from a conditioned aquarium can be added to the new filter bed. The gravel should come from an aquarium that has been maintained for several weeks at the same temperature as the new aquarium.

Water Replacement

Excess detritus impairs water circulation through a filter bed and is undesirable. As detritus accumulates, vertical channels form and the water follows these paths of least resistance, by-passing much of the filtrant. The result is erratic oxygen delivery throughout the filter and formation of anoxic areas where the growth of aerobic bacteria is inhibited. The use of very fine gravel or sand is also undesirable for the same reason, particularly in deep filter beds.

The removal of excess detritus entails stirring the surface of the filter bed and putting the material into suspension. Some can then be siphoned out along with 10% of the old water during each biweekly partial water change. The gravel should be stirred gently. In tests performed by Kawai et al. (1965), 1.0 g of surface from an old seawater aquarium filter bed was removed and washed gently in clean seawater. Afterward, 40% of the nitrifying ability of the sand was lost. Subsequent washings decreased it even more. When another gram was washed vigorously, 66% of the nitrifying capacity was lost, and this was reduced another 14% by a second washing. These findings show two things: (1) a considerable portion of the total number of nitrifiers is attached to detritus; and (2) vigorous washing detaches others from the gravel surfaces. The filter bed is a permanent installation. The gravel should never be taken out and washed. Washing detaches bacteria from gravel surfaces, in addition to removing too much of the detritus. In cases where it is absolutely essential to wash the gravel, do so directly in the aquarium using clean water of the same salinity. In seawater aquariums, clean seawater should be used; in brackish water and freshwater aquariums, use clean brackish water and aged tap water, respectively.

Nitrification is affected adversely by fluctuations in temperature, pH, dissolved oxygen, and salinity. Of these, temperature appears to be the most significant, followed by salinity. Changes in pH and dissolved oxygen are perhaps least important. Keep in mind that the animals and plants ordinarily kept in aquariums are far more sensitive than bacteria to physical and chemical changes in the water. Proper aquarium management thus dictates that the comparatively narrow tolerance limits of these higher organisms be considered before the needs of the filter bed biota, some cells of which usually manage to survive under the poorest conditions. Thus the guidelines established here, although quite narrow for most bacteria, are not too restrictive for other living things.

Shifts in environmental conditions should be anticipated both on a daily basis and before making partial water changes or replacing water lost by evaporation. Dissolved oxygen must be near saturation at all times, including its concentration in any makeup water. This is mainly for the benefit of the animals, because nitrifying bacteria are not highly aerobic. An acceptable pH range in brackish water and seawater is 8.0–8.3; a range of 7.1–7.8 is adequate for freshwater. The pH of the makeup water should match the value in the aquarium as closely as possible.

Temperature has the most noticeable effect on nitrification, and perhaps on mineralization processes too. Time lags in the conversion of nutrients can be expected if the water temperature is lowered. Temperature increases ordinarily accelerate bacteriological activities. Many ectothermic (cold-blooded) animals cannot tolerate even small temperature shifts without suffering ill effects, if such shifts are too rapid (Section 5-2). Changes in temperature that occur either naturally or following a partial water change or water addition should not exceed $1°C\ 24\ hr^{-1}$. This means that replacement water must be heated or refrigerated before use, depending on the situation. If an aquarium is at room temperature, water used for replacement or addition is best kept in the same room in a closed container and not used until its temperature equilibrates with the temperature of the aquarium.

Seawater used for partial water changes should match the salinity of the aquarium. In brackish water aquariums, dilutions of seawater should be made in a separate container and the replacement water added a little at a time. Allow the aquarium a short adjustment period between increments. Mixing makeup water separately relegates mistakes to the mixing container and does not jeopardize the stable salinity of the aquarium. Moreover, diluting seawater first minimizes the chance of salinity fluctuations in an already diluted medium.

Aged tap water is best to use for replacement of water lost by evaporation in aquariums of all salinities, and for partial water changes in freshwater aquariums. *Aged tap water* is tap water that has been held for 3 days in an open container and aerated to expel the chlorine. In brackish water and seawater aquariums, surface evaporation causes an increase in salinity, which, if gradual, has few adverse effects on most organisms. However, allowing large increases in salinity to occur before adding freshwater is a poor practice, particularly if the difference afterward is large. Brackish water and seawater aquariums should be kept covered to reduce the rate of evaporation, and freshwater should be added when the salinity increases by 0.2 ‰.

2

Mechanical Filtration

Mechanical filtration is the method used to lower turbidity in aquarium water. Turbidity can be caused by periodic "blooms" of algae or bacteria, but ordinarily it results from the appearance of particulate matter in the water column. Particulate matter is categorized by size and origin. *Particles* are the smallest bits of particulate matter; when they coalesce they become *aggregates*. Aggregates, in turn, clump together to become *detritus*. Detritus is the macroscopic floc or tan-colored "dust" commonly seen on the surfaces of old filter beds. The composition of detritus is complex, consisting of both inorganic and organic substances, the latter including living microorganisms. In this discussion, particles, aggregates, detritus, and free-floating algae and bacteria are referred to collectively as particulate organic carbon (POC), except where separate treatment is warranted.

Mechanical filtration separates and concentrates POC by trapping it against suitable substrata or septa. In a water treatment scheme, mechanical filtration follows biological filtration, but precedes physical adsorption and disinfection processes, as illustrated by Fig. 2-1. Biological and mechanical filtration occur simultaneously in subgravel filters, and design parameters are the same for both functions (Section 1.5).

2.1 PARTICLE, AGGREGATE, AND DETRITUS FORMATION

Baylor et al. (1962) and Sutcliffe et al. (1963) showed that particles in seawater could be produced by the adsorption of dissolved organic carbon (DOC) onto the surfaces of air bubbles. Batoosingh et al. (1969) discovered that small particles in the size range of 0.22–1.2 μm were important as "seed matter," or nuclei, in the formation of aggregates from small particles during bubbling. They also found that any particles, living or dead, accelerated aggregate formation, and there was no direct evidence that living bacteria were necessary to carry out the process, as had been postulated earlier by Barber (1966).

Batoosingh et al. (1969) determined that fine bubbles were more effective producers of aggregate material than large bubbles. This can be attributed

21

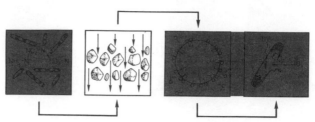

Figure 2-1. Position of mechanical filtration in a water management scheme. Original.

to the larger surface area of small bubbles for adsorption of surface-active organic matter from solution. Sheldon et al. (1967) found that the quantity of particles formed in filtered seawater samples could be increased by raising the concentration of DOC. When the level was doubled to 4.0 mg C l^{-1} by addition of glycine, there was a tenfold increase in particle formation.

Batoosingh et al. (1969) noticed that when bubbling rates were varied, the yield of aggregates was rapid at first, then tapered off so that a 24-hr experiment seldom produced more than twice as much material as one lasting only 3–6 hr. The presence of particles, either natural or artificially produced, seemed to inhibit further aggregate formation. Filtration on a continuous basis produced greater aggregate formation than intermittent filtration. Thus particles seemed to be necessary to initiate the appearance of aggregates during bubbling, but too many were inhibiting.

Aggregate and detritus formation in aquarium water probably occur by a similar process. Production probably takes place at two locations: (*1*) inside airlift pumps; and (*2*) at the air-water interface near the effluent from the airlift. Perhaps aggregates and detritus are formed on rising air bubbles in airlift pumps by a process analogous to the one postulated by Barber (1966) to occur in the sea. Adsorption of dissolved surface-active organics could take place on rising air bubbles. At the surface of the water, the ejected bubbles would fall back as collapsed monomolecular films, ready to adsorb other material. If this theory is correct, aggregate and detritus production in aquarium water of moderately high DOC concentrations can be accelerated by: (*1*) the use of smaller air bubbles in airlift pumps; (*2*) increased intensity of bubbling; and (*3*) continuous removal of newly formed particles by mechanical filtration. Procedure 3 would reduce inhibition of additional aggregate formation by the presence of too many particles. It appears that accelerated aggregate and detritus formation might be a useful tool for reducing the level of DOC in aquarium water indirectly. (Physical adsorption methods remove DOC directly from solution, as discussed in Chapter 3.) This could be accomplished by filtering out the detritus, along with the DOC adsorbed in its formation.

2.2 SUBGRAVEL FILTRATION

Subgravel filters remove POC from circulating water by entrapment against the surfaces of the gravel grains, or in the interstices among grains. Experts disagree on the exact means by which POC is separated from water by a bed of gravel. Some of the mechanisms thought to be important are straining, sedimentation, diffusion, and strong chemical bonding between the surface of the filtrant material and POC (Craft 1966, O'Meilia and Stumm 1967, Tchobanoglous 1970). The removal of POC by three of these mechanisms is shown diagrammatically in Fig. 2-2.

The efficiency by which POC is removed from water passing through a biological filter depends mainly on three factors: (*1*) grain size and distribution of the gravel; (*2*) shape of the grains; and (*3*) the presence of detritus.

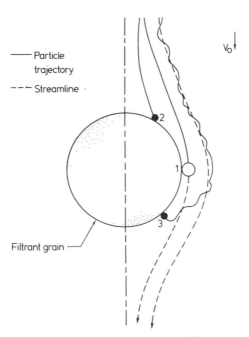

Figure 2-2. Entrapment of POC against a gravel grain by (*1*) straining, (*2*) sedimentation, and (*3*) diffusion. Redrawn from Yao et al. (1971).

Size and Distribution

The mechanical filtering efficiency of gravel increases with decreasing grain size. Small grains have more surface area on which to collect POC. Moreover, the small interstices facilitate removal of finer substances. This results in a greater percentage of suspended matter extracted per volume of filtered water.

Only one grade of gravel should be used in airlifted biological filters not fitted with backwash equipment. When different grades, or sizes, of gravel are mixed, the number of surfaces is reduced. Large voids form in the gravel bed in areas where large grains predominate. Detritus can work deeper into the bed in these places, becoming more difficult to remove.

The path that water takes in its downward movement through a filter bed depends on localized patterns of resistance. The flow is distorted when gravel is unevenly distributed on the filter plate. Thin parts of the bed offer less resistance than thick parts and attract a greater percentage of circulating water. This condition may lead to the same chronic turbidity problems often seen in aquariums with filter beds composed of ungraded gravel.

Shape

Rough, angular gravels are best for mechanical filtration. Their numerous surfaces increase the amount of aggregates and detritus that can be removed.

Detritus

The accumulation of some detritus in a filter bed enhances its mechanical filtering efficiency. Detritus fills in some of the interstices among gravel grains, enabling the filter bed to trap smaller bits of material. Old filter beds ordinarily produce greater water clarity then new filters because they contain more detritus.

2.3 RAPID SAND FILTRATION

Rapid sand filters are used in public aquariums, fish hatcheries, and aquaculture installations to process large volumes of influent water. Prefiltration of the water before use removes turbidity and many disease-producing organisms. At some installations, filters of this type are also used to recirculate culture water, especially if the animal load is high. Rapid sand filters are powered by mechanical pumps instead of airlifts, and have flow rates that are several times faster than airlifted subgravel filters. The degree of water clarity is no greater than can be attained with properly managed subgravel filters, but the more rapid flow rate and use of finer filtrant material reduces turbidity in less time.

The design and operation of rapid sand filters differ considerably from that of subgravel filters, although the mechanisms of POC removal (straining, sedimentation, and so forth) are identical. Surface area is not so critical, for one thing, because of the high flow rate. Moreover, removal of POC takes place throughout a greater depth of the filter bed instead of only at the surface.

One or more layers of graded silica gravel topped by a layer of fine sand are commonly used in rapid sand filters. The gravel is arranged in layers of increasing grain size from top to bottom. Some designs incorporate anthracite coal as the surface filtrant, with sand underneath, followed by three grades of silica gravel of increasing grain size. The combination of anthracite and gravel is referred to as *dual media*. The grain size of the anthracite is larger than the grain size of the sand, which allows a greater size range of POC to penetrate into the bed. The effect is to increase the surface area of the filter and prolong the interval between backwashes.

Gravity Sand Filters

Gravity sand filters are rapid sand filters with open tops, as shown in Fig. 2-3. Most are made of concrete and are very large. Circulation through the filtrant is sustained by removing water from underneath the bed. This creates a partial vacuum at the bottom, and the flow of water from the top by gravity is increased. The principle of water circulation is similar to that in subgravel filters, except that mechanical pumps are used and the flow rate is much greater. This type of filter is commonly used in drinking water and wastewater filtration. Most of the designs given in engineering texts can be adapted to large aquarium installations.

Pressure Sand Filters

The flow of water through a pressure sand filter is illustrated in Fig. 2-4. Water enters the closed pressure vessel through a baffle on the inside near the top. From there it is forced, under pressure, down through the filtrant. Pressure sand filters are manufactured in many sizes, mostly from steel. Units made of fiberglass-reinforced plastic are available from Jacuzzi

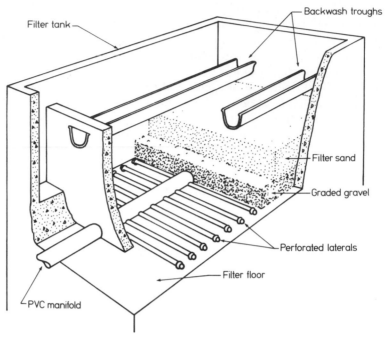

Figure 2-3. Cutaway view of a rapid sand gravity filter. Courtesy American Water Works Association.

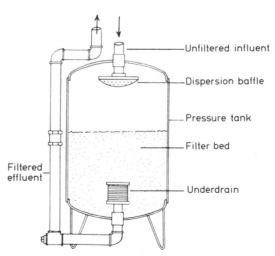

Figure 2-4. Cutaway view of a rapid sand pressure filter. Courtesy Paragon Swimming Pool Co., Inc.

Figure 2-5. Small rapid sand pressure filter made of fiberglass-reinforced plastic for filtering aquarium tanks of moderate size (about 4000 l). Courtesy Jacuzzi Bros. Inc.

Bros., Inc.* They are particularly well suited for processing seawater because none of the parts will corrode. A model that can be installed on aquarium tanks of moderate size (about 4000 l) is shown in Fig. 2-5.

2.4 DIATOMACEOUS EARTH FILTRATION

In diatomaceous earth (DE) filtration, a layer of graded skeletal remains of diatoms, held against a porous sleeve by vacuum or pressure, is used to strain POC from water. Diatomaceous earth filters are capable of removing smaller bits of POC than either subgravel or rapid sand filters. Some of the finer grades of DE remove bits of particulate matter as small as 0.1 μm, but not consistently.

The filtering mechanism of a DE filter consists of two parts. The first is a porous *central core*. The second part is the *filter sleeve*. The sleeve is preferably a thin, flexible, tightly woven polypropylene cloth that fits over the central core. Other materials, such as nylon, are less durable. The sleeves are removable in larger filter units. The central core and sleeve

*Jacuzzi Bros., Inc., 11511 New Benton Hwy., Little Rock, Ark. 72203

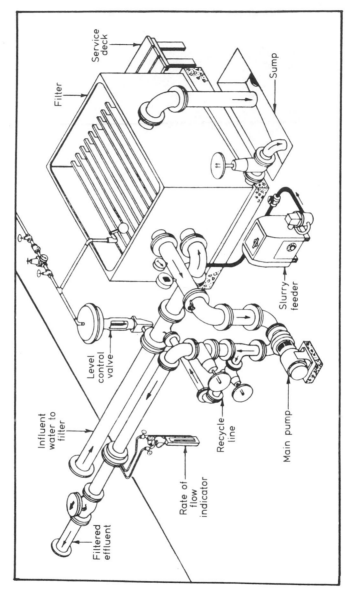

Figure 2-6. A DE filter installation for large water systems showing the filter bay, plumbing, slurry feed unit, and leaf-type elements. Courtesy BIF, General Signal Corp.

28

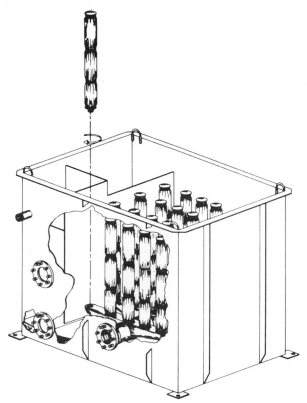

Figure 2-7. A DE filter for large water systems showing the filter bay, manifold, and column-type elements. Courtesy Keene Corp.

together make up the *filter element*. The sleeve supports a layer of diatomaceous earth; the central core supports the sleeve and exposes it to the circulating water. An arrangement showing a "leaf-type" filter element and plumbing is depicted in Fig. 2-6. The central cores cannot be seen because they are covered with sleeves. An arrangement with "column-type" elements and a manifold is shown in Fig. 2-7.

The layer of diatomaceous earth protects the sleeve from becoming clogged with POC and losing its porosity. This layer, called a *filter cake*, does the actual filtering. The principle is illustrated diagrammatically in Fig. 2-8. The figure shows a cross-sectional view through a sleeve and filter cake. Water passes through the filter cake and element, leaving suspended POC trapped in the DE.

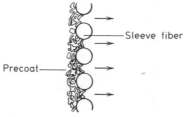

Figure 2-8. Diagrammatic cross-section showing the sleeve fibers and beginning of a filter cake, or precoat. Courtesy Johns-Manville Products Corp.

Gravity DE Filters

Designs for typical gravity DE filters are shown in Figs. 2-6 and 2-7. Filtering is accomplished in an open tank or *bay*. Water enters the bay and passes through the filter elements, the manifold, and into the pump. The principles of water flow are the same as those in gravity sand filtration. All gravity DE filters are designed to process large volumes of water.

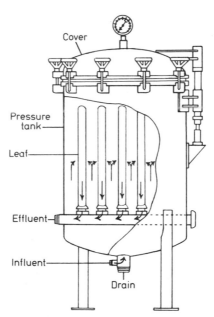

Figure 2-9. A DE pressure filter for large water systems showing the pressure tank, gauge, elements, and direction of water flow. Courtesy T. Shriver and Co.

Pressure DE Filters

A pressure DE filter for processing large volumes of water is shown in Fig. 2-9. The filter elements are sealed inside a pressure vessel. Water containing POC is forced through the filter elements, where POC is trapped in the filter cake, and the processed water is returned to the aquarium. Thus the operating principles are the same as those involved in pressure sand filtration. Pressure DE filters are manufactured in many sizes, including one for filtering small volumes of aquarium water (Fig. 2-10). These units are sold by Vortex Innerspace Products, Inc.,* and they work equally well in freshwater or seawater.

Efficient filtration with either gravity or pressure DE filters depends largely on three factors: (*1*) surface area of the filter elements; (*2*) precoat; and (*3*) slurry feed.

Figure 2-10. Diatomaceous earth filters for processing small volumes of aquarium water. Courtesy Vortex Innerspace Products Corp.

*Vortex Innerspace Products, Inc., 3317 East Bristol Rd., Flint, Mich. 48507.

Surface Area

The total surface area of the filter elements is important if the filter is to handle the amount of POC generated in the water system. A filter that is undersized requires frequent backwashes, which is time-consuming and expensive because of the DE that is discarded.

Precoat

After a DE filter is backwashed, it is coated with a new layer of diatomaceous earth called a *precoat* (Fig. 2-8). Previous to this, the old filter cake is flushed to waste along with accumulated POC. A new filter cake must be started to protect the sleeves initially from being coated by organic matter. First the water in the filter must be isolated from water in the aquarium and recycled as a separate closed system. This is accomplished with an arrangement of valves. To precoat, DE is added and recycling continues until the precoat becomes attached to the filter elements. Water containing newly added DE is milky. Precoating is complete when the water turns clear. Normal filtration is resumed by diverting the recycling water back to the aquarium. New (unused) culture water should be used to precoat because it is lower in DOC and POC.

No special precoat equipment is necessary for gravity filters. Dry DE is simply sprinkled onto the surface of the water while the filter is being recycled. Permanently installed DE pressure filters require a *precoat pot*. The correct amount of dry DE is mixed with water to form a slurry. This is added to the precoat pot, which funnels it into the pressure tank and onto the filter elements. A typical arrangement is shown in Fig. 2-11. The correct amount of precoat is 1.0 kg m^{-2} (dry DE) of filter sleeve surface area.

The plumbing in all permanent DE units must be designed so that the water flow does not have to be shut off after precoating to begin filtering the water. In gravity filters, DE is held against the elements by a partial vacuum inside the central core; in pressure units it is held in place by pressure applied directly to the filter cake. In both cases, the filter cake starts to fall off when the pumps stop.

Slurry Feed

The precoat is not adequate to maintain the porosity of the filter cake because of surface coating by POC. Most forms of POC are compressible and restrict water flow through the filter elements. In permanent units, filter cake porosity is maintained by adding small amounts of comparatively

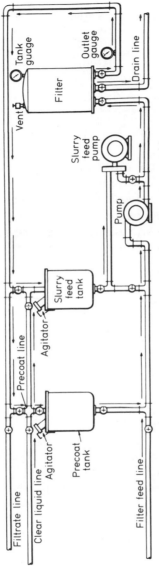

Figure 2-11. A typical DE filter installation for large water systems showing the precoat pot, plumbing, pressure filter, and slurry feeder. Courtesy Johns-Manville Products Corp.

33

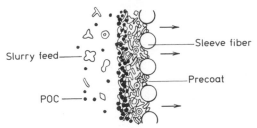

Figure 2-12. Diagrammatic cross-section of a filter sleeve showing the sleeve fibers and filter cake porosity maintained with slurry feed. Courtesy Johns-Manville Products Corp.

incompressible DE continuously to the outer surface of the precoat through a metering pump. These measured amounts of DE constitute the *slurry feed* (also called body feed). The principle is illustrated diagrammatically in Fig. 2-12. Without a slurry feed, the intervals between backwashes are shortened considerably. The right amount of slurry feed is determined by trial and error, but as a start, use 0.5 kg (dry DE) per square metre of filter sleeve surface area per hour.

2.5 MANAGEMENT PRACTICES

Rapid sand filters and DE filters must be cleaned periodically by backwash, and rapid sand filters can be used to remove excess detritus from subgravel filter beds. Diatomaceous earth filters need more maintenance than rapid sand filters, although filter selection depends on specific needs and on cost.

Cleaning Rapid Sand Filters

Rapid sand filters are cleaned by *backwash*, or reversing the flow of water through the filter bed. As the backwash water rises, gravel grains are lifted and the entire bed expands. Detritus, being lighter than the top layer of filtrant, rises higher and is forced out the overflow to waste. The filtrant grains are kept in suspension momentarily because the upward force of the water and the settling force of the grains are equal. After backwash, the filtrant settles back into the original graded layers (some intermixing occurs where layers adjoin). Trapped detritus is removed by three mechanisms during backwash. First, loose material is simply lifted from the surface of the filter bed and flushed out the overflow. Second, detritus deep in the bed and material adhering loosely to the filtrant is flushed out by the direct scouring action of the water. The larger grades of gravel in the lower layers disperse the backwash water evenly through the upper layers. Third, a portion of the detritus clinging to the filtrant grains is abraded off as the grains are lifted and collide.

Small pressure sand filters, such as the one depicted in Fig. 2-5, are useful for intermittent cleaning of larger subgravel filters. At installations away from the sea, the expense of artificial seawater often prohibits detritus removal by discarding large volumes of turbid water. A better arrangement utilizes a subgravel filter with an attached auxiliary pressure sand filter. The influent to this unit can be installed just above the surface of the filter bed of the subgravel filter. When the bed is stirred, suspended detritus is pulled into the pressure sand unit and clean water is returned to the aquarium.

Cleaning DE Filters

Most large DE filters are cleaned by backwash, although some gravity units are cleaned by draining the filter bay and washing away the filter cake with a strong jet of water from a garden hose. During a backwash, the flow of water is reversed and directed back into the manifold and through the central cores. This expands the sleeves and washes off the filter cake, which is then flushed to waste, as shown diagrammatically in Fig. 2-13. Backwashing should be done with tap water, even if the filter is used to process seawater. This prevents fouling the sleeves with organic matter and reducing the efficiency of the unit.

Filter sleeves should be taken off the central cores periodically and laundered to remove organic matter from the fibers. This can be done in a washing machine with a mild detergent and a water softener. Polypropylene sleeves are sensitive to high temperature and should never be washed in hot water. Excessive heat melts the material and closes the interstices in the cloth. Use warm or cold water only. After the sleeves are

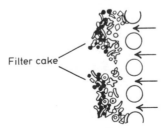
Filter cake

Figure 2-13. Diagrammatic cross-section of a filter sleeve showing the sleeve fibers and filter cake removal by backwash. Courtesy Johns-Manville Products Corp.

laundered, they should be rinsed several times in clean tap water to remove all traces of detergent.

Filter sleeves eventually become clogged. The signs of clogging are (*1*) a reduced interval between backwashes; and (*2*) bare spots on the sleeves after precoating. Clogging is caused by any of several factors, including the build-up of organic matter, precipitation of iron and manganous oxides, accumulation of carbonate scale, and algal growths. Ordinarily, the problem can be traced to organic material on the sleeve fibers, which can be removed by laundering. In seawater and hard freshwaters (groundwaters in particular), oxides and carbonate scale may be the cause. Algae are troublesome either in freshwater or seawater if the water is exposed to direct sunlight and heavily laden with nitrate and phosphate. To determine the cause of clogged filter sleeves and to correct the problem, use the following key after the filters have been drained and cleaned.

(*1*) Fill a pipette with orthotolidine or muriatic acid and squirt it on a clogged spot on an element. Let stand for 5 min, then rinse with tap water.

    a. The clogging substance does not change color, but the cloth turns white . . . . CA OR MG CARBONATE. Use *treatment 1*

    b. The clogging substance turns red; after rinsing, the cloth is white . . . . .IRON OXIDE. Use *treatment 1*

    c. The clogging substance turns dark gray or black . . . .(*2*)

    d. None of the above . . . .(*3*)

(*2*) Dissolve several crystals of sodium sulfite in a few millilitres of orthotolidine or muriatic acid and squirt it on the gray spot. The gray disappears and the cloth turns white . . . . .MANGANESE. Use *treatment 2*

(*3*) *a.* The elements feel greasy . . . .(*4*)

    b. The elements do not feel greasy. Squirt a few drops of 25% sulfuric acid on the spot. The spot is white after rinsing . . . . LIGHT ORGANIC COATING. Use *treatment 3*

(*4*) If no positive results are attained at this point, the elements have a HEAVY ORGANIC COATING, as indicated by the greasy quality of the cloth. This is most often caused by bacteriological decomposition of fatty components in fish flesh and oil (if fish flesh is used as food) and in animal excreta, although it is sometimes caused by algae. If the problem is recurrent, it can usually be traced to one of four conditions: improper maintenance of the filters (e.g., careless backwashing which leaves the elements dirty, or failure to use *treatment 3* routinely every 3 months), inadequate prefiltration with sand, insufficient physical adsorption, or backwashing or precoating with old aquarium water . . . . .Use *treatment 4*

The following treatments should be used in conjunction with the above key. Before starting treatment, the filter should be drained and the elements cleaned thoroughly and inspected. Column elements should be backwashed twice. Both column and leaf elements should be hosed off with a strong jet of tap water from a garden hose (gravity filters). It is also necessary to know the exact volume of the filter. The chemical additives used in the various treatments are given per 115 l of water.

Treatment 1

1. Fill the filter with clean tap water. Do not precoat.
2. Be sure the return valve to the aquarium is closed.
3. Add 1.0 l of 34% muriatic acid per 115 l of water in the filter to make a 0.25% solution.
4. Recycle for 15 min, or until the cloth turns white. When a heavy concentration of iron oxide is present, the solution will turn brown.
5. Drain the filter and backwash the elements (or hose off leaf elements) three times in succession, letting the unit drain completely after each backwash or hosing.
6. Fill the filter with clean freshwater, brackish water, or seawater. Precoat and resume normal filtration. The vacuum or pressure gauge should indicate zero after precoating. No bare spots should be visible on the elements.

Treatment 2

1. Fill the filter with clean tap water. Do not precoat.
2. Be sure the return valve to the aquarium is closed.
3. Add the same quantity of muriatic acid as in *treatment 1*.
4. Gradually add 9.0 g of sodium bisulfite per 115 l of water in the unit.
5. Recycle until the cloth turns white.
6. Drain the unit and backwash (or hose) the elements three times in succession, letting the filter drain completely after each cleaning.
7. Fill the filter with clean freshwater, brackish water, or seawater. Precoat and resume normal filtration. The vacuum or pressure gauge should indicate zero after precoating. No bare spots should be visible on the elements.

Caution: Do not use this treatment without adequate ventilation.

Treatment 3

1. Fill the filter with clean tap water. Do not precoat.
2. Be sure the return valve to the aquarium is closed.
3. Adjust the pH of the water to 5.0 with muriatic acid.
4. Add 0.38 l of 15% sodium hypochlorite per 115 l of water in the unit.

5. Recycle for 3 hr.

6. Reduce after 3 hr by adding 28 g of sodium thiosulfate per 115 l of water.

7. Continue to recycle until the chlorine residual (see tests in American Water Works Association et al. 1976) is 0.0 mg l^{-1} for three consecutive tests spaced 15 min apart.

8. Drain the unit and backwash (or hose) the elements three times in succession, letting the filter drain completely after each cleaning.

9. Fill the filter with clean freshwater, brackish water, or seawater. Precoat and resume normal filtration. The vacuum or pressure gauge should indicate zero after precoating. No bare spots should be visible on the elements.

Note: It is particularly important that the elements be as clean as possible before starting this treatment. Otherwise, the chlorine demand of the water may exceed the quantity of free chlorine in solution and result in incomplete oxidation of the organic film on the elements.

Treatment 4

1. Fill the unit with clean tap water. Do not precoat.

2. Be sure the return valve to the aquarium is closed.

3. Add 225 g of Calgon® and 112 g of laundry detergent per 115 l of water in the filter.

4. Recycle for 1 hr.

5. Drain the filter and backwash (or hose) the elements three times in succession, letting the unit drain completely after each cleaning.

6. Fill the filter with clean tap water and recycle for 30 min, then drain and backwash (or hose) once more.

7. Fill the filter with clean freshwater, brackish, or seawater. Precoat and resume normal filtration. The vacuum or pressure gauge should indicate zero after precoating No bare spots should be visible on the elements.

Note: Extra backwashing is necessary to remove all traces of residual detergent from the elements.

Filter Selection

The choice between a rapid sand or DE filter depends on specific needs and operating costs. Diatomaceous earth filters give superior water clarity, but they clog easily, require frequent maintenance, and are expensive to operate. The expense is incurred when a DE filter is operated continuously, because the filter cake must be discarded at the end of every filter cycle and replaced. On a yearly basis, the cost of operating a large DE filter can be considerable.

The operating costs can be reduced by connecting the DE filter in series after a rapid sand filter. This arrangement minimizes the POC load on the filter cake and prolongs the intervals between backwashes. The same arrangement is especially useful in prefiltering large volumes of raw influent water, which are often very turbid. Diatomaceous earth filters cannot be used to process influent water directly because the filter sleeves clog quickly and require frequent backwashes. The same factor limits the use of DE units as adjuncts in the cleaning of subgravel filters. On the positive side, DE filters provide better water clarity than rapid sand filters because they are able to extract smaller bits of suspended POC. Diatomaceous earth is the method of choice in situations requiring minimum turbidity.

Rapid sand filters are preferred if the purpose is (*1*) auxiliary cleaning of subgravel filters or (*2*) auxiliary biological filtration. Pressure-type units work better in the first situation because they are cheaper to install and take up less space. Gravity sand filters are superior in the second situation because of their greater surface areas. A rapid sand filter of either type will sustain working populations of bacteria so long as the unit is in continuous operation and not dried out part of the time. No DE filter, by comparison, is of any value in biological filtration because the filter cake is transient and never present long enough to make a significant contribution.

CHAPTER 3

Physical Adsorption

Dissolved organic carbon (DOC) can be removed from aquarium water by physical adsorption using either granular activated carbon or foam fractionation. The inorganic nutrients ammonia, nitrate, and phosphate can be reduced in solution by the use of ion exchangers. Ion exchange is ordinarily not considered a physical adsorption process, but it has been included here for convenience. Granular activated carbon and foam fractionation work well in freshwater, brackish water, and seawater; ion exchangers are useful only in freshwater. Physical adsorption follows biological and mechanical filtration in a water management scheme, and precedes disinfection, as illustrated in Fig. 3-1. Ion exchangers should follow the other physical adsorption methods in series.

Adsorption is defined as the collection of DOC on a suitable interface. The interface is the boundary between two phases, one being the aquarium water. The other phase can be a gas, such as an air bubble in a foam fractionation contactor, or a solid, such as an activated carbon grain. During adsorption, the substance collected (called the *adsorbate*) leaves the water and becomes bound to the surface of the gas or solid (called the *adsorbent*). If the chemical bond is strong, the adsorption process is irreversible and chemical adsorption is said to have occurred. If the bonds are weak, such as those formed by van der Waals forces, the process is one of *physical adsorption*. A weak bond means that the adsorption process is reversible, often as the result of a change in adsorbate concentration. *Desorption* is the reverse phenomenon by which adsorbates that have been adsorbed pass back across the phase boundary and reenter solution.

3.1 ACTIVATED CARBON

Some of the DOC can be removed from aquarium water by adsorption on activated carbon. Activated carbon is manufactured in two steps. The first is char formation in which a carbonaceous material such as coal, animal bone, wood, or nutshell is heated to a red heat (about 600°C) to drive off the hydrocarbons. Char formation must be done in the virtual absence of

40

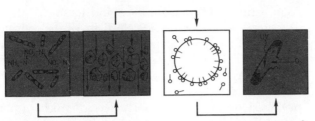

Figure 3-1. Position of physical adsorption in a water management scheme. Original.

air to prevent combustion. The second step is activation. The char is reheated, this time to about 900°C in the presence of an oxidizing gas. The gas develops the porous internal structures of the carbon, which then become the surfaces on which DOC is collected from water, as shown in Fig. 3-2. The size of the pores that are formed is not usually important in the removal of adsorbates from a liquid phase (Tchobanoglous 1972).

An activated carbon is either powdered or granular. Although powdered varieties have more surface area, they are expensive and difficult to handle. Emphasis here is on the use of granular types. Granular activated carbon (GAC) is defined as an activated carbon with a particle diameter greater than 0.1 mm (Tchobanoglous 1972).

Factors Affecting Adsorption Rate

Several factors limit the rate at which DOC is adsorbed by GAC. The most important are (1) mass transfer of the adsorbate into the activated

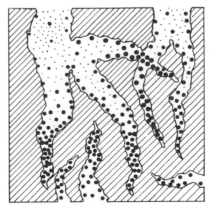

Figure 3-2. Diagrammatic cross-section of an activated carbon grain showing DOC adsorbed in the pores. Redrawn courtesy of Barnebey-Cheney.

carbon; (2) contact time; (3) concentration and nature of the adsorbate; (4) particle size, pore surface area, and selectivity of the activated carbon; and (5) the presence of biological slime on the surfaces of the GAC grains. Temperature and pH cannot be considered factors because they must be controlled within narrow limits to keep animals and plants alive. Of the two, pH is probably more important. Morris and Weber (1964) noted that adsorption equilibria ordinarily are not affected much by temperature changes, particularly over the range encountered in natural waters.

Tchobanoglous (1972) divided the adsorption process into three distinct steps: transfer of the adsorbate through the film of water and biological slime that surrounds the adsorbent; diffusion through the pores of the GAC; and formation of chemical bonds between the adsorbate molecules and the activated carbon. Only the first two steps are rate-limiting; the third takes place instantly. Thus the mass (molecular) transfer of a DOC molecule is controlled by the first two steps, as shown in Fig. 3-3.

Contact time between water containing DOC and activated carbon grains is critical. If the contact time is too short, mass transfer cannot occur. Contact time can be increased by lengthening the contact column or decreasing the flow rate of water passing through it.

Adsorption rate varies in part as the square root of the concentration of the adsorbate (Morris and Weber 1964). Therefore, a greater percentage of adsorbate is taken up per unit time in dilute solution and the process favors removal of trace concentrations of solutes. Nevertheless, the rate of adsorption increases with increasing concentration of solute, but the rela-

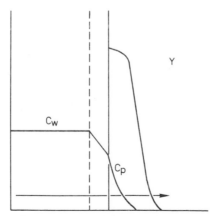

Figure 3-3. Combined film and pore diffusion model for adsorption of DOC by activated carbon: c_w = DOC concentration in the water, c_p = DOC concentration in the GAC, and y = DOC concentration in the pore water. Redrawn from Sontheimer 1974.

tionship is nonlinear. Morris and Weber also found that molecules of high molecular weight were taken up more slowly than smaller species. Moreover, the configuration of a molecule was a factor that affected how quickly it was adsorbed. Species that were highly branched were removed more slowly than others of similar molecular weight but structurally more compact.

Adsorption rate varies in part as the square of the diameter of individual carbon grains (Morris and Weber 1964). Once gathered at the water-GAC interface, diffusion of an organic molecule is rate-limited by mass transfer into the activated carbon grain, as mentioned previously. The particle size of a GAC grain also affects the rate of mass transfer, and uptake of DOC is faster on smaller grains. However, breaking up large grains of GAC into smaller ones, although it opens additional pores, does not increase the adsorption rate appreciably (Morris and Weber 1964).

The pore surface area of GAC can be measured in general terms by molasses, phenol, and iodine numbers. Each of these chemical compounds has a different molecular diameter and the extent to which it can be adsorbed is a function of how many pores of that particular size are available on the surfaces of the activated carbon. Iodine, as the smallest of these compounds, gives an indication of the total pore surface area. Molasses number gives an accounting of the largest pores, and phenol number describes the pores intermediate in size between iodine and molasses numbers. Iodine, molasses, and phenol numbers give an idea of the potential adsorptive capacity of a brand of GAC, but are not wholly reliable. Sontheimer (1974) found that in three brands of GAC tested to remove TOC (total organic carbon) from wastewater, neither the phenol number nor the surface area of the grains had any relationship to adsorptive capacity.

The last factor that affects adsorption rate is biological slime. GAC grains immersed in water soon acquire a coating of biological slime that is formed by bacteria. The material produces two effects: it interferes with mass transfer of DOC molecules into the grains (McCreary and Snoeyink 1977), and it accounts for continued removal of DOC after the physical adsorptive capacity of the GAC has been exhausted (Maqsood and Benedek 1977). The latter process is caused by the activities of heterotrophic bacteria present in the slime.

Design

GAC is best kept in a separate container, or *contactor*. Scattering it on the surface of a filter bed is impractical, because once the material is saturated it must then be separated from the gravel grains and removed. In small aquariums, standard corner filters powered by airlifts make suit-

able contactors. Tight plugs of glass wool should be placed on top of the
GAC to reduce coating of the grains with colloidal material.

GAC contactors for larger aquariums can be constructed easily from a
length of PVC pipe, as illustrated in Fig. 3-4. Each end of the contactor
should be threaded and fitted with removable caps for easier maintenance.
The effluent end should have an inset perforated plate, a section of plastic
flyscreen, or a plug of glass wool to prevent the GAC from being sucked
into the airlift. The end caps are drilled, tapped, and fitted with threaded
flexible hose nipples. The design should incorporate a by-pass arrangement, like the one shown in Fig. 3-5, to allow water to recirculate through
the biological filter when the contactor is shut down for recharging. Ideally,
a GAC contactor should follow the biological filter in series, as illustrated
in Fig. 3-6. Heterotrophic oxidation mineralizes many of the biodegradable
fractions of the DOC beforehand, which reduces the organic load on the
activated carbon and prolongs its life.

The GAC contactor shown in Figs. 3-4 and 3-6 is recharged by diverting
the flow of water directly back to the aquarium, then unscrewing the
influent and effluent caps. This step is easier if flexible hose is used at the
connection points instead of rigid PVC pipe. The exhausted GAC must

Figure 3-4. Contactor for GAC made from a length of PVC pipe. Courtesy Mystic
Marinelife Aquarium.

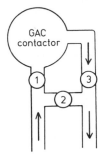

Figure 3-5. By-pass arrangement for a GAC contactor. Valves *1* and *3* are open and *2* is closed for normal filtration. Valve *2* is opened and *1* and *3* are closed to recycle through the biological filter while the GAC is being changed. Original.

then be replaced with new material that has been washed in clean tap water to remove the dust. The threaded male ends of the fittings can be wrapped with one thickness of Teflon® tape to prevent leaking.

A GAC contactor for still larger aquariums can be made from a 200-l (55 U. S. gal) steel drum with a removable lid, as illustrated in Fig. 3-7.

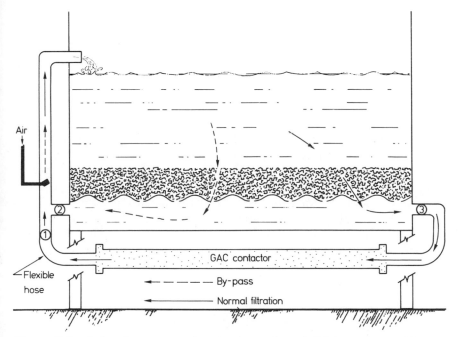

Figure 3-6. A closed-system aquarium showing tank stand, GAC contactor, and by-pass valves. The valve arrangement is the same as shown in Fig. 3-5. Original.

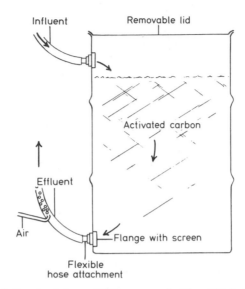

Figure 3-7.   Steel drum GAC contactor holding 200 l. Original.

The inside of the drum and lid should first be painted with two coats of a durable epoxy paint to retard rusting. Two 2.54-cm holes are then drilled in the side of the drum, one near the top and the other at the bottom. Threaded PVC flanges (2.54 cm diameter) are sealed against each hole on the inside with silicone sealant and stainless steel bolts, and PVC flexible hose attachments are screwed into the threaded flanges from the outside. A subgravel plate to suspend the GAC above the bottom is unnecessary if a small section of plastic flyscreen is attached to the inside face of the flange. This is adequate to keep the GAC from being carried into the airlift returning to the aquarium. Influent water to the contactor should come from a by-pass in the return line to the biological filter, as depicted in Fig. 3-5. The drum should be filled three-fourths full with washed GAC. One drum should be used for every 4000 l of aquarium water.

Very large water systems require GAC contactors equipped with mechanical pumps. Most pressure sand filters are suitable. These units are filled with GAC instead of sand and gravel. The best designs include removable tops, as shown in Fig. 3-8, because the activated carbon must be replaced periodically.

3.2   FOAM FRACTIONATION

Many surface-active fractions of the DOC can be concentrated and removed in foam produced by foam fractionation. This process is also called

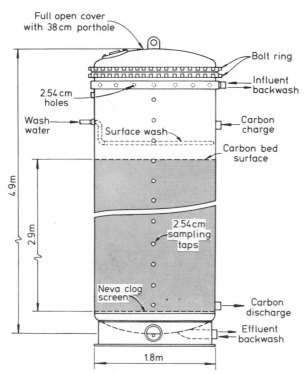

Figure 3-8. A pressure GAC contactor for very large water systems. Adapted from Parkhurst et al. (1967).

airstripping and protein skimming; the latter is inaccurate in implying that only proteinaceous substances are removed. According to Rubin et al. (1963), foam fractionation removes DOC by two mechanisms: (*1*) surface-active materials are adsorbed physically at the surfaces of rising air bubbles in a contact column; and (*2*) nonsurface-active compounds may combine chemically with surface-active material and be collected. Foam fractionation contactors (or simply foam fractionators) also remove some of the particulate organic carbon by entrapment in the foam and are an auxiliary mechanical filtration technique. Microorganisms in suspension comprise a portion of the POC, and Schlesner and Rheinheimer (1974) demonstrated that foam fractionation reduced the number of free-floating microorganisms in aquarium water.

When immersed in water, a surfactant molecule is polar at one end and nonpolar at the other. The polar end, if attracted to water molecules around it, becomes hydrophilic, whereas the nonpolar portion is not attracted to water molecules and is hydrophobic. As the name suggests, surface-active molecules tend to congregate near the surface with their

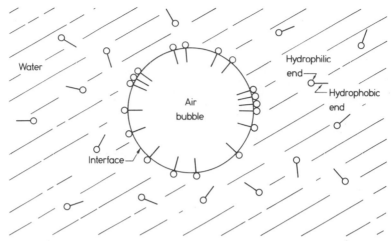

Figure 3-9. Diagrammatic illustration of an air bubble adsorbing the surface-active ends of DOC molecules onto its surfaces. Redrawn from Ng and Mueller (1975).

hydrophilic ends submerged and their hydrophobic parts in contact with the air, as shown diagrammatically in Fig. 3-9. This behavior allows them to be concentrated at the interface of rising air bubbles.

Foam fractionation does not reduce the level of ammonia in aquarium water, as is commonly thought, because the pH cannot be manipulated without endangering animals and plants. Kuhn (1956) showed that significant amounts of ammonia (as NH_3) were not removed in airstripping towers used to treat wastewater until the pH was raised to 11.0.

Factors Affecting Foam Fractionation

The two main factors that affect the efficiency of foam fractionation are (1) contact time between air bubbles and the water and (2) bubble size. Contact time, in turn, depends on the flow rate of water entering and leaving the contact column, column height, and air volume. Contact time is increased by any method that prolongs the retention of air bubbles in the contact column. The two most common techniques are lengthening the column or decreasing the flow rate of influent water. Small bubbles have a greater surface-to-volume ratio than large bubbles and are preferable.

Design

Three foam fractionator designs for use in aquariums are shown in Figs. 3-10 and 3-11. The two in Fig. 3-11 can be scaled up for use in large water

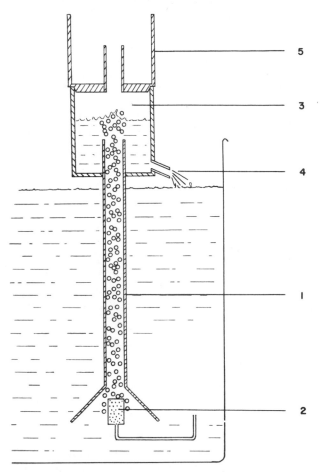

Figure 3-10. A cocurrent foam fractionator. Numbers are explained in the text. Redrawn from Sander (1967).

systems. Figure 3-10 depicts the *cocurrent* method, in which air bubbles rise in the contact column with the flow of water. The devices in Fig. 3-11 use the *countercurrent* method, where air bubbles rise against the downward flow of water. The countercurrent design assures greater contact time because the water flowing in the opposite direction has a "ballasting" effect on the bubbles and slows their rate of ascent.

In the cocurrent design, air from a compressor is injected through a diffuser (*2*). As the bubbles rise (*1*) they mix, or "contact," with the water. The surface-active fraction of DOC is adsorbed onto the bubble surfaces, producing a foam at the air-water interface (*3*). As foam accumulates, it

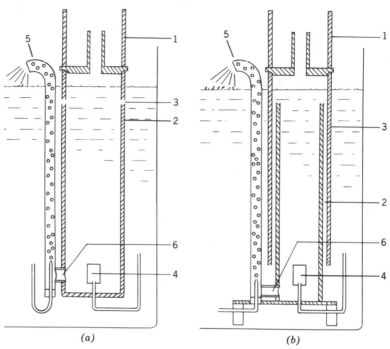

(a)   *(b)*

Figure 3-11. Two countercurrent foam fractionators. Numbers are explained in the text. Redrawn from Sander (1967).

spills into a collection chamber (*5*). The collection chamber can be removed and cleaned. Treated water is returned to the aquarium at (*4*).

In the countercurrent design, water moves downward against rising air bubbles. In Fig. 3-11*a*, air moves from a compressor through a diffuser (*4*) and into the contact column (*2*). Untreated aquarium water enters the contact column near the top (*3*). In this design, the top of the column (*2*) also serves as a separation chamber, and excess foam passes into a collection chamber at (*1*), which can be removed for cleaning. Treated water is removed from below the surface instead of from above, as in the cocurrent design. Water passes from a connecting tube (*6*) near the bottom of the contact column and is airlifted back to the aquarium (*5*). In Fig. 3-11*b*, the mechanism is similar except that the contact column (*2*) is fitted with a larger outside column (*3*) that forms a sheath. The advantage is that untreated water in the contact column cannot be driven back to the aquarium by injected air, as it can be at (*3*) in the design shown in Fig. 3-10*a*. This makes design 3-10*b* slightly more efficient.

3.3   ION EXCHANGE

Ion exchange materials composed of natural zeolites or synthetic resins are effective in removing several undesirable contaminants from aquarium water. The use of the ion exchange process is mainly limited to freshwater, because competition from other ions in brackish water and seawater reduce the number of exchange sites available to contaminant species. Ion exchange as a water management process has been largely ignored by aquarists. This is surprising, considering that under good working conditions ion exchange removes more than 90% of the ammonium ion, nitrate, and phosphate from solution.

Ion exchangers can be defined as electrochemically charged zeolites or resin beads that remove an ionic species from solution by exchanging it for another species of similar ionic charge. According to Kunin (1963), ion exchange resins are manufactured with the following properties: strongly acidic cation, weakly acidic cation, strongly basic anion, and weakly basic anion.

Ammonium Ion Removal

The removal of ammonium ion from wastewater using ion exchange was discussed by Nesselson (1954), Culp and Slechta (1966), Battelle Memorial Institute (1969), Koon and Kaufman (1971, 1975), and Jorgensen et al. (1976). Jorgensen et al. (1976) studied NH_4^+ extraction with clinoptilolite, a natural zeolite. The material was found to work by a combination of ion exchange and adsorption. The saturation point was about 8.0 m-mole NH_4-N 100 g^{-1}. Adsorption and ion exchange increased with increasing contact time up to 120 min.

Spotte (1970) suggested that ion exchangers could be used to remove nutrient ions, including ammonium ion, from aquarium and hatchery water. The efficacy of the process was confirmed by Johnson and Sieburth (1974) and Konikoff (1974). Johnson and Sieburth found that both clinoptilolite and an ion exchange resin lowered the level of ammonia (total NH_4-N) in freshwater from a salmon hatchery. Konikoff noted that when new freshwater aquariums were stocked with high densities of catfish, those filtered with clinoptilolite columns showed much lower ammonia concentrations than the controls, which were equipped with conventional biological filters. Subsequent increases in nitrite, however, resulted in high mortalities in both types of aquariums. Water in the control aquariums showed gradual increases in ammonia, nitrate, phosphate, acidity, and specific conductivity, and a decrease in alkalinity and pH. Aquariums filtered with clinoptilolite had lower acidity, phosphate, and ammonia levels compared with the controls. Nitrate and specific conductivity were similar to the controls, and alkalinity and pH were higher.

Specific studies on removal of nitrate and phosphate from aquarium and hatchery water have not been made, but some of the factors controlling the extraction of these substances from wastewater are mentioned in the following sections.

Factors Affecting Ion Exchange

Ion exchange is limited by (*1*) interionic competition; (*2*) fouling of the exchange sites by DOC; and (*3*) particle size of the exchanger.

Interionic Competition

The efficiency with which ammonium ion is removed is a function of the cationic strength of the influent water; in other words, the ionic strength when only the cationic species are considered. Cationic strength can be calculated by

$$I_+ = \tfrac{1}{2} \sum (m_i z_i^2) \tag{7}$$

where m_i is the cation concentration of the ith species in moles 1^{-1}, and z_i is the valence of the cation of interest (Koon and Kaufman 1975). In studies with wastewater, Koon and Kaufman (1975) found that the ammonium ion exchange capacity of clinoptilolite decreased sharply with increasing cationic strength up to about 0.01 moles 1^{-1}. Further increases in the cationic strength of the water continued to diminish the exchange capacity, but to a lesser degree. Johnson and Sieburth (1974) also noted that the efficiency with which clinoptilolite removed ammonium ion from salmon hatchery water declined sharply when the ionic strength of the water was increased. At a salinity of only 5 ‰, total NH_4-N removal dropped tenfold (Fig. 3-12). Salinities of 10, 15, and 25 ‰ reduced efficiency even more, but not so dramatically as the initial decrease that occurred at 5 ‰.

The removal of nitrate and phosphate is similarly impaired in waters of high ionic strength. Eliassen et al. (1965) noted that a high chloride concentration in the influent waste water reduced the number of exchange sites available for contact with nitrate and phosphate. High sulfate had an inhibiting effect as well. Buelow et al. (1975), studying nitrate extraction from waters of different ionic strengths, showed that the selectivity of ion exchange resins could be changed drastically by altering the total anion concentration in the influent water. In some cases, preferences for the species removed were reversed. In water of low ionic strength, sulfate was adsorbed preferentially to nitrate, but this was reversed when the ionic

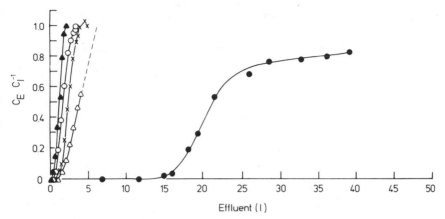

Figure 3-12. Adsorption of ammonia by an ion exchange resin from distilled water (●); and artificial seawater at salinities of 5‰ (△); 10‰ (×); 15‰ (○); and 25‰ (▲). C_E and C_I = effluent and influent concentrations of ammonia, respectively. Redrawn from Johnson and Sieburth (1974).

strength of the influent water was increased. The results of all these studies follow a typical pattern. In general, both cation and anion exchangers show marked preference for polyvalent ions in solutions of weak ionic strength,

Buelow et al. (1975) found that silica in the influent water coated the surfaces of ion exchange resins, preventing nitrate exchange, rather than by replacing nitrate as the species removed. The most serious interfering anion was sulfate. Bicarbonate and chloride ions were less serious. In dilute solutions, these ions were removed in the order sulfate > nitrate > chloride > alkalinity (the sum of carbonate and bicarbonate). In stronger solutions, the positions of sulfate and nitrate were reversed.

Species selectivity—and therefore efficiency—of the ion exchange process is mainly a function of interionic competition. Selectivity decreases sharply with increasing ionic strength of the influent water. From a practical standpoint, ion exchange is a low-yield process in seawater, brackish water, and even hard freshwater. Jorgensen et al. (1976) reported that ammonium ion removal was far less effective in wastewater or tap water than in distilled water because of increased competition from calcium ions. Similarly, Eliassen et al. (1965) wrote that a chloride content of only 200 mg l^{-1} and sulfate at 65 mg l^{-1} seriously impeded removal of nitrate and phosphate from wastewater by ion exchange. Considering that calcium, chloride, and sulfate occur in seawater at concentrations of 400, 1.9×10^4, and 885 mg l^{-1}, respectively, it is easy to understand why available sites on ion exchangers become occupied quickly and the material becomes exhausted before significant quantities of nutrients can be extracted. Davey et al. (1970) and Schiøtz (1976) reported that ion exchange resins removed trace metals from seawater, which is undoubtedly true, but how this was done selectively was not mentioned.

DOC Fouling

The presence of DOC has an adverse effect on inorganic ion exchange. Organic compounds in solution foul the surfaces of ion exchangers and reduce the number of sites available for adsorption of inorganic species. This was reported by Jorgensen et al. (1976) and Frisch and Kunin (1960). Evidence of organic fouling during phosphate removal from wastewater was discussed by Eliassen et al. (1965) and the effects are shown graphically in Fig. 3-13. The fouling substances could be removed with sodium hy-

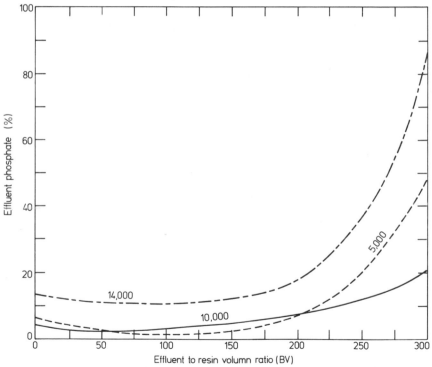

Figure 3-13. Average removal of phosphate from wastewater by ion exchange. The upward shift of the curve indicates decreased efficiency in phosphate removal. The shift to the left indicates reduced adsorptive capacity of the resin caused by DOC fouling of the exchange sites. Numbers on the curves represent continuous flow in bed volumes (BV). Redrawn from Eliassen et al. (1965).

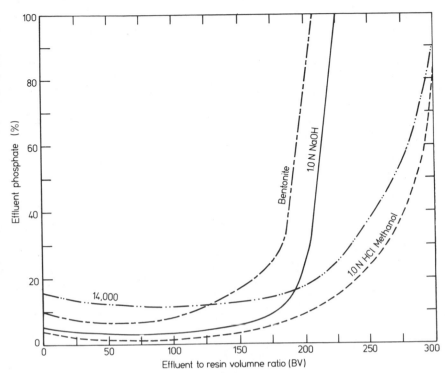

Figure 3-14. Regeneration of ion exchange resins by various methods. Numbers on the curves represent continuous flow in bed volumes (BV). Redrawn from Eliassen et al. (1965).

droxide, hydrochloric acid, and methanol, and by backwashing the resin column with bentonite, which acted as a mechanical scrubber at the resin surfaces. Bentonite was the most effective method, as seen in Fig. 3-14. These authors suggested that DOC coats the resin grains instead of undergoing direct ion exchange, forming monomolecular films. Johnson and Sieburth (1974) noted that the ability of clinoptilolite to adsorb ammonium ion was exhausted in the presence of 30.0 mg TOC l^{-1}. As a result, the material was only 50% as efficient in removing 95% of the ammonia from distilled water, as shown in Fig. 3-15.

Grain Size

Grain size of the exchange material influences adsorptive capacity. Smaller grains have more surface area for exchange with the water. Jorgensen et al. (1976), working with an ion exchange resin in ammonium

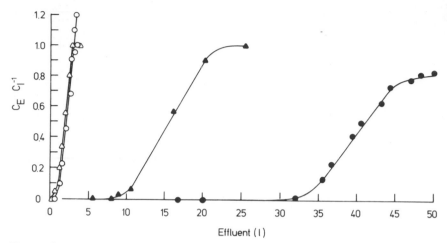

Figure 3-15. Adsorption of ammonia by clinoptilolite (18 × 45 mesh) from distilled water (●), fish culture water containing 30 mg DOC l^{-1} (▲), artificial seawater at salinity 15‰ (○), and fish culture water at salinity 15‰ (△). C_E and C_I = effluent and influent concentrations of ammonia, respectively. Redrawn from Johnson and Sieburth (1974).

ion removal from wastewater, noted that at a grain size of 2.5–5.0 mm, the static capacity was 0.47 meq NH_4-N g^{-1}; at 1.4–2.0 mm, static capacity was 0.62 meq NH_4-N g^{-1}. Johnson and Sieburth (1974) also reported that larger grains of ion exchange material were less efficient in ammonium ion removal. Their work showed that the ideal mesh size for clinoptilolite grains was 18 × 45 (1.00 × 0.35 mm).

Design

The design of ion exchange contactors is considered in two parts: (*1*) selection of the ion exchange material; and (*2*) the ion exchange contactor.

Selection of Ion Exchangers

There are no general guidelines for predicting how a specific ion exchanger will perform under given conditions. Buelow et al. (1975) emphasized that pilot studies should be performed using the actual water to be treated, instead of distilled or simulated waters. They noted that the selectivity and competition of a given ion exchange material may be altered or even reversed by changes in the ionic strength of the influent water. At high ionic strength, nitrate is adsorbed preferentially to sulfate; the reverse is true when the ionic strength of the influent water is lowered. The removal of ammonium ion continues even as the ionic strength increases. Overall, removal efficiency is improved when interionic competition is minimized, regardless of the contaminant species removed. The general affinities of ion exchangers for the three major inorganic contaminants are summarized in Table 3-1, but the information is presented only as a guide and should not be considered a substitute for pilot studies.

The selection of an ion exchanger should be based on the chemical nature of the contaminant removed and the ionic strength of the water to be treated. Anionic species like nitrate and phosphate require anionic exchangers; cationic species, such as ammonium ion, are extracted from solution by positively charged materials. Weaker ion exchangers are preferable to strong ones if they do the job, because they are easier to regenerate. In certain cases, materials that remove large quantities of contaminant in a single pass are not always the best choice. A more important consideration in aquarium operation is how well the material performs after successive regenerations. Buelow et al. (1975) remarked, for example, that ion exchange resins that removed high initial concentrations of nitrate were sometimes harder to regenerate.

Koon and Kaufman (1975) stated that clinoptilolite was perhaps the best material for ammonium ion removal, although the resin tested by Johnson and Sieburth (1974) was also effective. Strongly basic anion exchange resins in chloride form are suitable for nitrate removal from wastewater (Buelow et al. 1975, Nesselson 1954). Eliassen et al. (1965) removed 92% of the nitrate and 95% of the phosphate from wastewater with a strongly basic anion exchange resin, and Martinez (1962), using a resin of the same type but from a different manufacturer, achieved removals of 99% and 98% for nitrate and phosphate, respectively, from wastewater.

**Table 3-1. Types of Ion Exchangers
Suitable for Removal
of Contaminant Species**

Ion Exchanger	Contaminant Species Removed
Basic cationic	Ammonium ion
Basic anionic	Nitrate ion
	Phosphate ion
	Sulfate ion

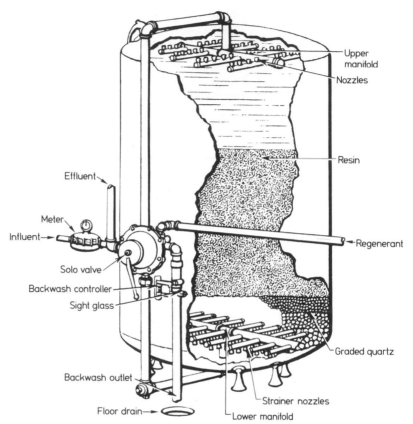

Figure 3-16. A rapid sand pressure filter filled with ion exchange resin beads for treating large volumes of water. The valve arrangement allows the resin to be regenerated in placed. Courtesy Rohm and Haas Company.

Contactors

Ion exchange materials can be placed in contactors of a size suited to the volume of water being treated. The designs shown in Figs. 3-4 and 3-6 for holding GAC are adequate. In very large water systems, a rapid sand pressure filter can be used to hold ion exchange material, as shown in Fig. 3-16.

The positioning of an ion exchange contactor is important. Jorgensen et al. (1976) suggested that ion exchange contactors be placed in series after GAC contactors to reduce fouling by DOC. The GAC contactor, in turn, should be preceded in series by an efficient mechanical filter. This reduces the POC level and prevents the surface layers of either the ion exchanger or the GAC from becoming clogged.

3.4   MANAGEMENT PRACTICES

All physical adsorption processes probably concentrate and remove trace ions from solution. This is not serious unless algae are cultured in the water, in which case biweekly 10% partial water changes become critical.

Activated Carbon

The adsorptive capacity of GAC is exhausted eventually and the material must be replaced or reactivated. If this is not done, desorption occurs and DOC leaves the pores of the adsorbent and reenters solution. Reactivation is impractical unless high heat and steam under pressure are used together, and even then only the largest pores are reopened (Joyce and Sukenik 1964). Granular activated carbon thus becomes less effective with each reactivation, and probably can never be restored to its original adsorptive capacity.

A workable rule of thumb is to use 1.0 g GAC per litre of aquarium water and replace the material every 2 months. Mineralization of DOC by heterotrophic bacteria attached to GAC grains sometimes gives the impression that the material is still viable when it is not (Maqsood and Benedek 1977). Continuous monitoring of the total organic carbon (TOC) in the effluent water from the GAC contactor is the only sure way of knowing when DOC "breakthrough" has occurred and organic substances are being desorbed.

Foam Fractionation

In airlifted foam fractionators, the volume of injected air is the same as would ordinarily be used in an airlift pump of similar dimensions (Table 5-1, p. 77). Percent submergence is also the same, and effluent water should emerge in a smooth stream. Contact time is erratic if the volume of air exceeds the capacity of the contact pipe. Adsorption is partly a function of bubble surface area and a diffuser should always be used at the air injection point.

Ion Exchange

Freshwaters with high iron and manganese concentrations must be pre-treated by settling, strong aeration, or oxidation with ozone or potassium permanganate. Buelow et al. (1975) found that iron precipitates reduced nitrate removal from wastewater by fouling the surfaces of the exchange resins. The precipitates could be removed by treating the resin with 1.0 N HCl and backwashing. Waters originating from wells (groundwaters) are characteristically high in both iron and manganese.

Ion exchange processes take place rapidly and are a function of the number of available exchange sites (volume of the ion exchange material) and flow rate. These factors can be determined experimentally in pilot studies for a given water supply, or arbitrarily by monitoring the length of time necessary for exhaustion of the ion exchanger to occur, then manipulating the flow rate and volume of material accordingly. Exhaustion takes place with the sudden appearance (breakthrough) of high concentrations of the nutrient species in the effluent water.

Table 3-2 can be used as a guide for regeneration of ion exchange resins. Most ion exchangers can be regenerated with sodium chloride solutions, and Korngold (1972) suggested that seawater could be used, which would

**Table 3-2.   Suggested Regeneration Levels
for Ion Exchange Resins.
(Courtesy Rohm and Haas Company)**

Ion Exchange Resin	Ionic Form	Regenerant	Requirement (meq Regenerant per meq Resin)
Strong acid cation	H^+	HCl	3–5
	H^+	H_2SO_4	3–5
	Na^+	NaCl	3–5
Weak acid cation	H^+	HCl	1.5–2
	H^+	H_2SO_4	1.5–2
	Na^+	NaOH	1.5–2
Strong base anion I	OH^-	NaOH	4–5
	Cl^-	NaCl	4–5
	Cl^-	HCl	4–5
	SO_4^{2-}	Na_2SO_4	4–5
	SO_4^{2-}	H_2SO_4	4–5
Strong base anion II	OH^-	NaOH	3–4
Weak base anion	Free base	NaOH	1.5–2
	Free base	NH_4OH	1.5–2
	Free base	Na_2CO_3	1.5–2
	Cl^-	HCl	1.5–2
	SO_4^{2-}	H_2SO_4	1.5–2

have obvious advantages in seaside installations that maintain freshwater organisms. As a rule of thumb, a 10% NaCl solution is a good starting point in determining the concentration of regenerant required, although each situation is different. Buelow et al. (1975) found, for example, that in waters of low ionic strength, regeneration of ion exchange resins was achieved with 0.359 kg m^{-3} (3.0 lb 1000 gal^{-1}) of NaCl, compared with 1.366 kg m^{-3} (11.4 lb 1000 gal^{-1}) after treating with water of higher ionic strength. When waters from two freshwater wells were used, the amounts of NaCl required were 0.605 and 0.623 kg m^{-3} (5.05 and 5.2 lb 1000 gal^{-1}). Ordinarily, most regeneration requirements can be determined by varying the volumes of regenerant.

Sodium chloride and the substances used by Eliassen et al. (1965) to remove fouling DOC (sodium hydroxide, hydrochloric acid, and methanol) may be directly toxic, or cause significant pH changes in poorly buffered water. Regenerants and cleaning compounds must be used with care. Ion exchange contactors should be removed from the water system for regeneration, if practical. Afterward, the contact column must be flushed liberally with tap water.

CHAPTER 4

Disinfection

Disinfection is the destruction of pathogenic organisms by the application of physical or chemical agents. Disinfection of aquarium water ordinarily is accomplished by ultraviolet (UV) irradiation or ozonation. Either process reduces the number of free-floating microorganisms, but has no effect on infectious or parasitic organisms attached to host animals. Neither UV light nor ozone produces reaction products with residual effects that are adequate to kill pathogens after initial contact. As such, disinfection may be useful in preventing reinfection, but is not a substitute for antibiotics or other chemotherapeutic agents in the treatment of diseases.

Disinfection is affected adversely by the presence of dissolved and particulate organic carbon (DOC and POC), and UV sterilizers or ozonators should be placed in series after the biological and mechanical filters and physical adsorption contactors, as illustrated in Fig. 4-1.

4.1 ULTRAVIOLET IRRADIATION

Ultraviolet radiation kills microorganisms in water directly by deactivating DNA (deoxyribonucleic acid) within the cells. UV lamps produce *radiation*, and the process by which water is treated is *irradiation*.

Factors Affecting Percent Kill

The effectiveness of a UV sterilizer depends on three factors: (*1*) size of the organism; (*2*) the amount of radiation generated; and (*3*) the penetration of UV rays into the water. In general, the larger the organism, the more resistant it is to UV. Many viruses, bacteria, and the smaller life stages of fungi and protozoans can be killed by irradiating them with 35,000 μW sec cm^{-2} (microwatt seconds per square centimetre of UV lamp). Larger organisms, such as the tomite (swarming) stage of the freshwater parasitic ciliate *Ichthyophthirius*, may require nearly 400,000 μW sec cm^{-2} (data in

62

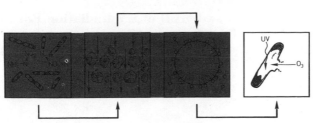

Figure 4-1. Position of disinfection equipment in a water management scheme. Original.

Hoffman 1974). The adult stage, which is attached to the skin and gills of the host fish, requires perhaps 1,717,200 μW sec cm^{-2}, if it can be detached (Vlasenko 1969). These differences in dosage levels reflect the great size disparity between the tomite and adult stages of the parasite (20 × 35 μm versus 800 μm). *Oodinium ocellatum,* a dinoflagellate parasitic on marine fishes, can probably be killed by 35,000 μW sec cm^{-2} at the free-swimming (dinospore) stage, but this has not been confirmed experimentally. The tomites of another troublesome parasite of marine aquarium fishes, the ciliate *Cryptocaryon irritans,* may require a minimum lethal dose (MLD) of about 800,000 μW sec cm^{-2}, but again this is only speculation. The assumption is based on its size (35 × 56 μm according to Nigrelli and Ruggieri 1966), compared with the tomites of *Ichthyophthirius,* for which MLD data are available. The MLD values for several microorganisms, most of them parasitic, are presented in Table 4-1.

UV rays probably cannot penetrate water farther than about 5 cm under ideal conditions. Penetrating power is further reduced by DOC, POC, and inorganic ions in solution. Thus UV irradiation is less effective in waters that have high concentrations of DOC or that are turbid. The presence of large concentrations of inorganic ions in brackish water and seawater means that UV irradiation is less effective than in freshwater at similar dosage levels.

Design

Wheaton (1977) provided a synopsis of UV equipment designs for aquaculture. There are two types of UV sterilizers, suspended and submerged. Suspended sterilizers consist of batteries of UV lamps and reflectors hung 10–20 cm above a shallow trough through which water flows. The trough is necessary to reduce the depth of the water that is being irradiated.

Submerged sterilizers are more reliable. Their operating parameters can be engineered more precisely and they can be installed at any point in a

Table 4-1. Sizes and MLD of UV Radiation For Some Microorganisms Free-Living or Parasitic in Aquarium or Hatchery Water. From Data in Hoffman (1974)

Microorganism	Life Stage	Size (μm)	MLD (μW sec cm^{-2})
Trichodina sp.	—	16×20	35,000
Trichodina nigra	—	22×70	159,000
Saprolegnia sp.	zoospore	4×12	35,000
Saprolegnia sp.	hypha	8×24	10,000
*Oodinium ocellatum**	dinospore	8×12	—
Sarcina lutea	—	1.5	26,400
Ichthyophthirius sp.	tomite	20×35	336,000
Ichthyophthirius sp.	tomite	20×35	100,000
Cryptocaryon irritans†	tomite	35×56.5	—
Chilodonella cyprini	—	35×70	1,008,400
Paramecium sp.	—	70×80	200,000

* From data in Nigrelli (1936).
† From data in Nigrelli and Ruggieri (1966).

water system simply by plumbing in the influent and effluent lines. In good sterilizers, the UV lamp is encased in a watertight quartz jacket and is not exposed directly to the water. The jacket provides an insulating layer of air that lets the lamp function near its optimum temperature (40.5°C)

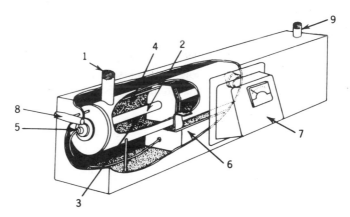

Figure 4-2. A UV sterilizer and component parts: (*1*) influent; (*2*) UV lamp; (*3*) jacket; (*4*) disinfection chamber; (*5*) lamp socket; (*6*) ballast; (*7*) intensity meter; (*8*) hand wiper; and (*9*) effluent. From Spotte (1973).

without suffering cooling effects from the water. Efficiency is impaired considerably if a jacket is not used and the lamp must operate at a lower than optimum temperature. A typical submerged UV sterilizer is shown in Fig. 4-2. Water enters the unit at (*1*), flows past the UV lamp and jacket at a fixed rate (*2*), and returns to the aquarium through the effluent line (*9*).

Two auxiliary devices are mandatory on submerged UV sterilizers: a UV intensity meter and an apparatus to wipe the jacket. UV lamps gradually become less efficient with time. The wiper is needed to keep biological slime from building up on the outside of the jacket. Accumulated slime reduces the penetrating power of the rays.

UV sterilizers should be capable of a maximum radiation output of 1.0 $\times$ 10^6 μW sec cm^{-2} at a flow rate of not less than the entire volume of the water system every 24 hr. Contact time is critical, and the effective dose of radiation can be doubled by reducing the flow rate through the sterilizer by one-half.

4.2 OZONATION

With the right equipment used properly, injection of triatomic oxygen (O_3), or ozone gas, into aquarium water reduces the number of free-floating microorganisms. At ordinary dosage levels and contact times ozonation does not lower the level of dissolved organic carbon by oxidizing it directly to free CO_2, as is sometimes thought, nor does it oxidize an appreciable amount of the total ammonia to higher oxidation states. Organic carbon compounds that are susceptible to ozonation are those with carbon-carbon double bonds. Functional groups attacked are $-SH$, ^+S, $-NH_2$, $-OH$, and $-CHO$. During ozonation, oxidized organics are merely converted to other dissolved organic compounds. This is easily demonstrated by the negligible change in TOC concentration of a treated wastewater effluent (Farooq et al. 1977a, Nebel et al. 1973). After ozonation, seawater at Mystic Marinelife Aquarium did not show any decline in TOC from values that averaged about 6.0 mg TOC l^{-1}.

Huibers et al. (1969), McCarthy and Smith (1974), and Nebel et al. (1973) concluded that ozonation is not a nitrification process. Singer and Zilli (1975) demonstrated that in freshwater over the pH range 7.0–9.0, slow kinetics precluded the oxidation of significant amounts of ammonia to nitrate. Similar results can be projected for seawater, although Honn and Chavin (1976) reported that ammonia and nitrite levels decreased and nitrate increased in a closed-system seawater aquarium after ozonation. The implication was that injection of ozone into the effluent from the biological filter oxidized ammonia and nitrite directly. However, no kinetic studies were performed, and direct oxidation of inorganic nitrogen by ozonation in seawater is still unproved.

Factors Affecting Percent Kill

Two factors control the efficiency of ozone as a disinfectant: (1) contact time; and (2) the residual concentration of undissociated O_3. Residual levels of ozone are difficult to maintain. Ozone is very unstable and its oxidizing power is easily exhausted on substances other than live microorganisms. Thus the factors that ultimately determine the percent kill of suspended microbes are those controlling the dissociation rate of O_3 after it is injected into a contact chamber.

These factors exert an ozone demand on aquarium water and lessen the amount of O_3 available for disinfection: DOC, POC (including microorganisms), pH, and the ionic strength of the water. As any one of them increases there is a concomitant rise in the threshold level of ozone needed for disinfection. Temperature also influences percent kill. Farooq et al. (1977b) reported that the percent kill of microorganisms was greater at high temperatures. At a constant ozone residual (0.57 mg O_3 1^{-1}) and four temperatures (9, 20, 30, and 37°C), they showed that disinfection of wastewater increased markedly with increasing temperature. The results are given in Fig. 4-3.

The dissociation rate of ozone accelerates with increasing pH. The effect of pH on disinfection is indirect, as reported by Farooq et al. (1977a), and the percent kill of microorganisms seems to be related more to the ozone residual than to pH per se. As shown in Fig. 4-4, the percent survival of the yeast *Mycobacterium fortuitum* was similar at four pH values and relatively constant concentrations of residual ozone.

The presence of DOC or POC increases ozone demand in the water and limits the number of microorganisms that can be killed. Hoigné and Bader (1976) reported that when microorganisms and DOC were present together, DOC was preferentially attacked. This occurred partly because organic molecules are not so widely dispersed as microorganisms, and because the dissociation products (free radicals) of ozone react more readily with DOC than does undissociated O_3 (Hoigné and Bader 1976). Farooq et al. (1977a) found that the cell density of suspended microorganisms was an important factor in the disinfection process. Lower disinfection rates were seen at high cell densities. Microorganisms must be considered as

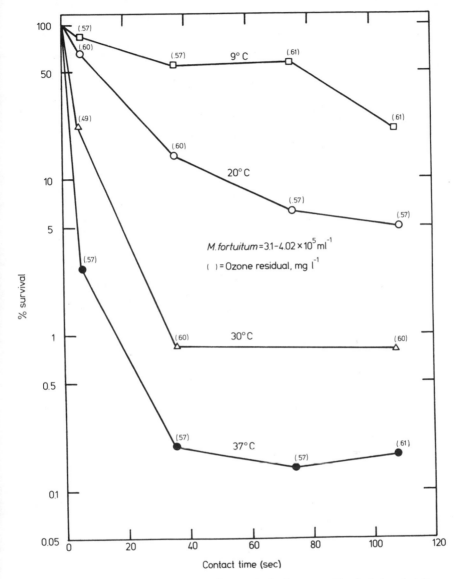

Figure 4-3. Effect of temperature on the survival of a yeast, *Mycobacterium fortuitum*, for a constant ozone residual at a given contact time. Redrawn from Farooq et al. (1977b).

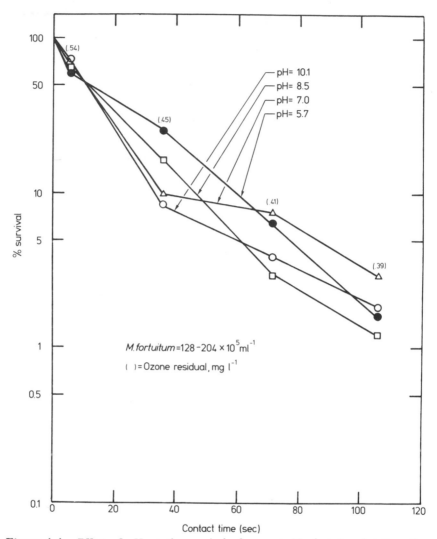

Figure 4-4. Effect of pH on the survival of a yeast, *Mycobacterium fortuitum,* for a constant ozone residual at a given contact time. From Farooq et al. (1977a).

particulate organic carbon for ozonation purposes, and POC in any form diminishes the disinfection potential of ozone.

An increase in the ionic strength of the water raises the ozone demand by increasing the number of inorganic ions susceptible to oxidation. Ozone residuals are more difficult to maintain in seawater than in freshwater because of the differences in the concentration of inorganic solutes.

Design

It is difficult to design ozonation systems for aquariums because so little information is available. Studies from the wastewater literature are sometimes applicable, provided that distilled water was not used in the experimental procedure. Few studies have been performed on the disinfection of seawater with ozone, and none in aquarium seawater. Honn et al. (1976) reported on the design of a silent discharge ozonator for use in closed-system seawater aquariums, but provided no data on disinfection. The susceptibility of fish and invertebrate pathogens to ozonation has been examined only sporadically. Nearly all work has dealt with bacteria, and protozoans and other infectious and parasitic organisms have been virtually ignored. Conrad et al. (1975) reported that ozonation reduced the numbers of the bacterium *Flexibacter columnaris* in water from a freshwater salmon hatchery, but they derived no conclusions about proper contact times or dosage levels. Wedemeyer and Nelson (1977) described effects of ozonated water on two bacterial fish pathogens, *Aeromonas salmonicida* and the enteric redmouth organism (ERM). Both are troublesome in salmonid hatcheries. In phosphate-buffered distilled water with no ozone demand, the ERM bacterium was eliminated by a residual of 0.01 mg O_3 l^{-1} within 0.5 min, and *A. salmonicida* within 10 min. To match the 0.5 min contact time for a 100 % kill, *A. salmonicida* required a dosage level of 0.04 mg O_3 l^{-1}. In soft and hard lake waters, a considerably higher dosage level (90 mg O_3 l^{-1} hr^{-1}) was needed to destroy both organisms within 10 min. Disinfection was more difficult in hard water than in soft water, as would be expected.

In terms of bacterial cell densities and TOC concentration, aquarium waters with low animal densities are comparable to good quality ground water, and those with high densities of animals are analogous to poor quality surface water. For treating such waters, McCarthy and Smith (1974) recommended dosage levels of 0.5–4.0 mg O_3 l^{-1} and contact times of 5–10 min. These parameters should be used as rules of thumb for freshwater, brackish water, and seawater aquariums until more specific information is available. The requirements are not too stringent. Keep in mind that parasitic protozoans are likely to be more resistant to ozonation than the smaller viruses and bacteria.

Two types of ozone generators are commonly used to disinfect aquarium water: (*1*) ultraviolet (UV), and (*2*) silent (electrical) discharge. An example of the second type is shown in Fig. 4-5. UV generators are often used where low ozone concentrations are required. The small ozonators designed for the home aquarium market ordinarily are of this type. If oxygen is used as the feed gas, UV generators produce concentrations ranging

Figure 4-5.   A small silent electric discharge ozone generator designed for laboratory use, but also suitable for ozonating small volumes of aquarium water. Courtesy PCI Ozone Corp.

from 1 to 10 mg O_3 l^{-1} (1.0 mg O_3 hr^{-1}), depending on the size of the unit. If an air feed is used, the output is at least 50% less. Feed gas flows either directly past the UV lamp or in an adjacent quartz jacket, depending on the design of the unit. In UV generators, the output (yield) of ozone is a function of the total effective radiation emitted in the range of 1000–2000 Å. (angstroms). This, in turn, is dependent on design of the UV lamp and its total emission area, lamp current, whether the feed gas is air or oxygen, feed gas pressure in the air space or discharge gap, and temperature of the feed gas.

Silent discharge generators are used where high yields of ozone are required, such as in public aquariums and fish hatcheries. Silent discharge ozonators can produce up to 6% wt of ozone, but most efficient production occurs at 1%–3% wt output. In simple terms, a silent discharge ozonator is an alternating voltage applied across two electrodes separated by an insulator, or dielectric, in the discharge gap (Fig. 4-6). The discharge gap must contain a dielectric for ozone production to occur; otherwise, there would be nothing but a spark or arc. In addition, the current must alternate because none of it can pass through the dielectric. The insulating material collects charges of electrons on its surface during a half-cycle of the alternating current and releases them when the polarity reverses.

When the ozonator is working, the discharge gap is filled with a diffused glow called the corona. The electrodes are usually made of stainless steel and aluminum and the dielectric is borosilicate glass. One electrode is contiguous with the glass dielectric; the discharge gap is located between

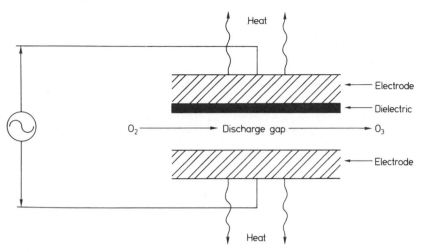

Figure 4-6. Basic ozonator configuration. Redrawn from Rosen (1973).

the dielectric and the second electrode, as seen in Fig. 4-6. The distance of the discharge gap at voltages less than 1500 V is commonly 1–3 mm. Silent discharge ozonators transform most of the input electrical energy into heat and some form of cooling is necessary unless the gas flow entering the discharge gap is high in proportion to electrode area. Too large a gas flow, however, reduces the efficiency of ozone production.

Ozone output of a silent discharge generator is a function of oxygen purity of the feed gas, feed gas temperature, feed gas flow, peak voltage, frequency, capacitance of the discharge gap, pressure of the feed gas in the discharge gap, and the capacitance of the dielectric. For a given ozonator configuration, ozone output is a function of current density when all other conditions are constant (Rosen 1973).

The requirements for dosage levels and contact times stated earlier in this section can be met easily in the treatment of large water systems, but are difficult to maintain in small volumes. There are two reasons for this. First, the amount of O_3 generated by UV ozonators is sufficient to disinfect only the tiniest water systems. Second, the available contactor designs for use in small aquariums are unsuitable. The foam fractionators pictured in Figs. 3-10 and 3-11 are typical. The design of these devices was first published by Sander (1967), who intended them to be used as ozone contactors, yet none retains water for the minimum recommended time of 5 min. Moreover, if ozone is generated in sufficient quantity to kill microorganisms, there is danger of enough residual O_3 remaining to be toxic to fishes and invertebrates. The fact that animals in small aquariums rarely are killed by ozonation attests to the general ineffectiveness of the equipment, and inadequate pretreatment of the water to lower ozone demand.

4.3 MANAGEMENT PRACTICES

UV irradiation is a more reliable process than ozonation. The level of radiation is easier to control than the quantity of ozone in solution, and UV equipment is inherently more efficient and can be engineered with more precision. Ozonators are notably inefficient, and most of the electric current applied to generate O_3 in silent discharge generators is dissipated as heat, which increases the cost of operation in large water systems. In addition, the actual concentrations of ozone generated must be confirmed *in situ* by laboratory analysis. Many commercial units are fitted with gauges to establish the flow rate and concentration of ozone gas, but because ozone generation is affected by so many factors, the gauges are unreliable. One unit tested at Mystic Marinelife Aquarium produced only 30% of the ozone shown on the gauge when the gas was subjected to analysis.

UV Irradiation

The longevity of a UV lamp depends on either the life of the electrode or solarization, or both. Electrode life declines with the number of times it is switched on and off. *Solarization* is the slow darkening of the inside of the lamp glass with continued exposure to UV radiation. Some of the mercury that has been vaporized condenses on the glass, filtering out an increasingly greater percentage of the UV rays being generated. A UV lamp does not "burn out" in the conventional sense, nor does it dim or flicker with prolonged use. The effects of solarization can be measured on the intensity meter. When the output has declined by 25%, the effective life of the lamp has been reached and it should be replaced. This normally occurs after about a year of continuous use, or 8000 hr.

It is questionable whether UV sterilizers reduce fish mortality in aquariums. Herald et al. (1970) reported that a submerged sterilizer installed on the water system at Steinhart Aquarium in San Francisco lowered the numbers of free-floating bacteria by 98%. Nevertheless, fish mortality remained the same (2.5% of the population per month). The question really revolves around how many organisms must be present to be infectious. The answer is not known. There is little doubt that UV sterilizers reduce fish mortality in hatcheries (Bullock and Stuckey 1977, Hoffman 1975, Sanders et al. 1972). The hatchery environment is quite different from a closed-system aquarium, however. The high animal densities characteristic of hatcheries may exacerbate disease transmission, and the presence of fewer host species increases the possibility that epizootic outbreaks of disease may occur.

Ozonation

It is critical that the feed gas be dried before being fed into an ozonator. Desiccators usually are preceded in series by a refrigerator for cooling the feed gas. Refrigeration does two things: (*1*) it eliminates some of the moisture in the feed gas, and this makes the desiccators more efficient; and (2) it reduces the temperature in the discharge gap, thus slowing down the decomposition of ozone molecules as they are produced. Water in the feed gas in amounts as low as $0.02—0.03$ mg H_2O l^{-1} seriously impairs ozone yield (O'Donovan 1965). Common desiccants used are silica gel and calcium chloride. These materials can be regenerated by passing a stream of hot air through them in the opposite direction of the normal feed gas flow.

Large water systems treated with a proper dosage level of O_3 require that the treated effluent be directed to an unoccupied chamber and aerated before being pumped back to the main water system. This is done to expel excess O_3 and O_2. Residual ozone is directly toxic to fishes and invertebrates, and the oxygen formed when ozone dissociates may supersaturate the water, causing gas bubble disease.

As in the case of UV irradiation, it is doubtful whether the injection of ozonated gas into aquarium water reduces the incidence of reinfection of aquatic animals to a significant extent. Further research is needed before the answer can be known.

CHAPTER 5

Gas Exchange
and Respiration

As used here, the term *gas exchange* refers to the diffusion of gases between air and water, and the mechanical means, such as aeration, by which it is made more efficient in aquariums. The physiological exchange of gases between aquatic organisms and the environment is referred to simply as *respiration* to avoid confusion with the first process.

5.1 GAS EXCHANGE

Animals and aerobic bacteria need oxygen to live, which in aquariums is supplied by aeration. In simple terms, the delivery of oxygen from the atmosphere to the water is a function of gas exchange between the water surface and the air, and between the surfaces of rising air bubbles and the water. The aeration process disrupts the surface of the water, and in doing so exposes a portion that is oxygen-depleted to the air-water interface. Because the concentration of O_2 in the air is greater, oxygen diffuses down the concentration gradient and enters solution. Much the same thing occurs when air bubbles injected beneath the surface are allowed to rise to the top. The bubbles contain a higher oxygen concentration than the water around them and lose O_2 to solution by diffusion. Aeration of an aquarium can be made more efficient if the surface of the water is disrupted continuously instead of intermittently, and if the air bubbles injected into the water are small. The smaller the bubble, the more surface area it has for gas exchange.

Surface agitation also drives free carbon dioxide from solution by bringing a greater portion of respired water to the air-water interface in a given period of time. This keeps the level of free CO_2 in close approximation with the quantity in the atmosphere by reducing its partial pressure in solution.

74

The Airlift Pump

Both surface disruption and bubble injection can be accomplished by airlift pumping. The main part of an *airlift pump* (or simply an airlift) is a vertical pipe called a *lift pipe,* part of which extends beneath the filter plate of the biological filter. Dissolved oxygen is most depleted underneath the filter plate because the water has just passed through the gravel bed, which contains a large population of aerobic bacteria. The rest of the lift pipe extends through the water column, and, normally, for a short distance above the surface of the water, as illustrated in Fig. 5-1.

An airlift pump is the most trouble-free means of moving water through a biological filter. The advantages of an airlift over mechanical pumps are lower initial cost, lower maintenance (an airlift has no moving parts), easy installation, portability, the fact that it is nonclogging, small space requirements, simplicity of design, easy construction, greater efficiency than centrifugal pumps when operating at low head and high submergence, easily regulated flow rate, and highly versatile application.

The operating principles of an airlift can be described as follows. When a pipe is submerged in water in a vertical position, the levels inside and outside the pipe equilibrate. Air is lighter than water and when air is injected at the lower end of the pipe it forms bubbles, which rise. As they

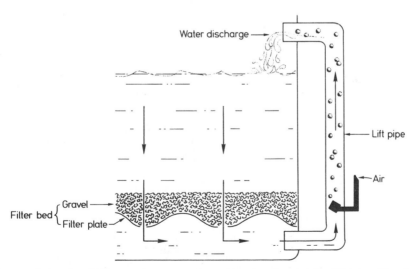

Figure 5-1. An airlift pumping arrangement showing the air inlet, lift pipe, filter bed, and movement of water. Original.

rise, the bubbles produce a mixture of air and water that is lighter than water alone. The air-water mixture inside the pipe is therefore lighter than the water outside and the equilibrium is upset. When this happens, heavier water from underneath the filter plate moves into the lower end of the pipe. So long as air is injected, equilibrium never takes place and the air-water mixture spills out from the top.

The main factor affecting the efficiency of an airlift is the *percent submergence* of the lift pipe. The volume of air necessary to operate an airlift increases with decreasing percent submergence. Maximum efficiency is attained at 100%, or when the top of the lift pipe is submerged. The minimum percent submergence acceptable is 80%.

Percent submergence is a simple calculation. In Fig. 5-2, if the distance between the air inlet and the discharge point (*3* to *1*) equals 50 cm and the total lift (*2* to *1*) is 10 cm, then

$$\frac{50\text{ cm} - 10\text{ cm}}{10\text{ cm}} = 80\% \text{ submergence} \qquad (8)$$

The flow rate for airlift pumps as a function of length and diameter using

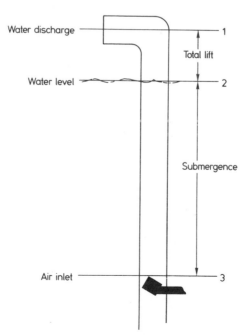

Figure 5-2. Operating principles of an airlift pump. Original.

eq. 10 at three values of submergence is given in Table 5-1. As a general rule, doubling the diameter of the lift pipe increases its capacity 5.6 times.

An airlift is less efficient when the volume of injected air exceeds the capacity of the lift pipe. This inefficiency is easily detected by the gurgling sound of air escaping through the top. It can be overcome by decreasing the air flow from the compressor. The effluent water should emerge in a smooth, even stream. If it spurts, the cause can usually be traced to one of two factors: either the air volume is too great for the diameter of the lift pipe and much of the air is escaping directly through the water and into the atmosphere, or the percent submergence of the lift pipe is not great enough.

As mentioned previously, gas exchange is more efficient if the air entering the lift pipe is diffused. In small aquariums where the diameter of

Table 5-1. Flow Rate, Q (1 min⁻¹), for Airlift Pumps as a Function of Length and Diameter Using Eq. 10 at Submergence Values of 0.8, 0.9, and 1.0. Courtesy Lee C. Eagleton and Gary Adams

Length, L (cm)		Lift Pipe Diameter, D (cm)					
		1.0	2.0	3.0	4.0	6.0	8.0
30	0.8	1.7	7.7	18.7	35.3	86.1	162.1
	0.9	2.0	9.2	22.4	42.2	102.9	193.7
	1.0	2.4	10.9	26.5	50.0	122.0	230.0
50	0.8	2.0	9.1	22.2	41.9	102.2	192.4
	0.9	2.4	10.9	26.6	50.0	122.1	229.8
	1.0	2.8	12.9	31.4	59.2	144.0	272.0
75	0.8	2.3	10.4	25.5	48.0	117.0	220.4
	0.9	2.7	12.5	30.4	57.3	139.8	263.2
	1.0	3.2	14.7	36.0	67.7	165.0	311.0
100	0.8	2.5	11.5	28.1	52.8	128.9	242.7
	0.9	3.0	13.7	33.5	63.1	153.9	289.8
	1.0	3.5	16.2	39.6	74.5	182.0	342.0
150	0.8	2.9	13.2	32.1	60.5	147.6	277.9
	0.9	3.4	15.7	38.4	72.2	176.2	331.9
	1.0	4.0	18.6	45.3	85.3	208.0	392.0
200	0.8	3.2	14.5	35.4	66.6	162.5	306.0
	0.9	3.8	17.3	42.2	79.5	194.0	365.0
	1.0	4.4	20.4	49.8	93.8	229.0	431.0
300	0.8	3.6	16.6	40.5	76.3	186.1	350.4
	0.9	4.3	19.8	48.4	91.0	222.2	418.3
	1.0	5.1	23.4	57.0	107.4	262.0	493.0

the lift pipe is 2.5 cm or less, an airstone is sufficient to disperse the air. In larger lift pipes, a plate containing many small perforations should be inserted above the point of air injection.

Design

Two basic airlift designs are used in aquarium applications: the *central airline* (airline inside the lift pipe), and the *side-air inlet* (airline outside the lift pipe). Either is suitable in most applications. Two variations of each type are illustrated in Fig. 5-3.

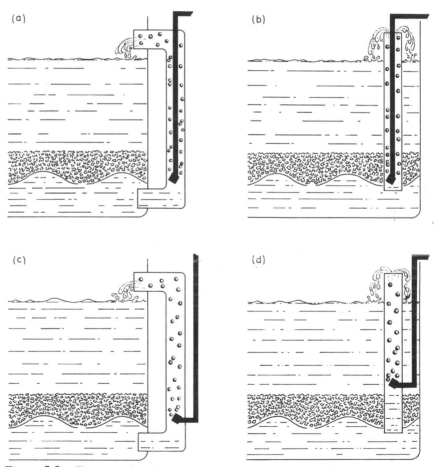

Figure 5-3. Two central airline designs (*a* and *b*), and two side air-inlet designs (*c* and *d*) for airlift pumps. Original.

Despite the many studies of airlifts through the years, a widely accepted equation relating the pertinent variables has yet to be developed. Nevertheless, sufficient information is available for most aquarium purposes. Castro et al. (1975) presented a report on the performance of airlift pumps of short length and small diameter. Német (1961) developed a correlation based on a large number of airlifts of various lengths, diameters, and submergence values. When the Német correlation is adapted to airlifts used for pumping water, the result is eq. 9:

$$Q = [0.504\,S^{3/2}L^{1/3} - 0.0752]D^{5/2} \qquad (9)$$

where Q = maximum flow rate of water when air flow is optimum (l min^{-1}), S = submergence (pipe length below the water divided by total length), L = pipe length (cm), and D = pipe diameter (cm).

Castro (1976, per. comm.) obtained about 140 sets of data over a range of $L, S,$ and D values appropriate for aquarium or aquaculture airlift pump design. Eagleton (1978, per. comm.) fitted most applicable sets of data to an equation of the form developed by Német. The flow rate dependence on diameter shown by the Castro data were fit more reliably by a 2.2 power on the diameter, rather than 2.5, as suggested by Német. Equation 10, which resulted from the linear regression, fits the Castro data well enough for design purposes, although possibly a better equation could be developed:

$$Q = [0.758\,S^{3/2}L^{1/3} + 0.01196]D^{2.2} \qquad (10)$$

The Castro data (Castro 1976, per. comm.; Castro et al. 1975) used to obtain eq. 10 covered a range of diameters from 1.7 to 7.8 cm, lengths from 60 to 300 cm, and submergence values from 0.6 to 1.0 (60%–100% submergence).

The Castro data were obtained for pumping water in a range of variables of likely interest to aquarists, whereas Német used data from many sources, including those reporting on oil flow from oil wells. As such, the Castro data and eq. 10 are recommended here. Table 5-1 gives the flow rate obtained, assuming that air flow rate is optimum, for a number of different airlift configurations using eq. 10. If the calculations presented in the tables were repeated using eq. 9, it would be seen that the Német correlation gives lower flow rates than found by Castro for 1–2 cm diameter pipes, and larger flow rates for pipes of 6–8 cm in diameter. The equations give similar results for pipe diameters of 3–4 cm.

A safety factor of about 25% is recommended when using eq. 10 for airlift pump design (Eagleton 1978, per. comm.). In other words, length, diameter, and submergence should be selected to give a calculated flow rate that is about 25% larger than needed. This factor guards against problems related to obtaining the optimum flow rate, and also provides a margin to cover minor restrictions to maximum flow, either at the bottom or the top of the pipe. A sample airlift design problem follows.

Design an airlift pump for an aquarium tank 90 cm deep and with a surface area of 3.0 m^2.

1. Determine the water flow rate using the rule of thumb that each 1.0 m^2 of surface area requires 40 l min^{-1} of flow rate (this is equivalent to 0.7 × 10^{-3} m sec^{-1}, or 1.0 gal ft^{-2} min^{-1}). Therefore,

$$(3.0 \text{ m}^2)(40 \text{ l m}^{-2} \text{ min}^{-1}) = 120 \text{ l min}^{-1}$$

$$\text{plus } 25\% = 150 \text{ l min}^{-1}$$

2. Choose a submergence value (e.g., 0.9). From Table 5-1, find the pipe diameter needed for the flow rate determined in the first step. Thus $S = 0.9$ and $L = $ depth/submergence $= 90/0.9 = 100$ cm, and $D = 6.0$ cm. Equation 10 can be solved for D (eq. 11), and D can be calculated directly, but ordinarily this is not necessary

$$D = \left[\frac{150}{0.758(0.9)^{1.5}(90)^{1/3} + 0.01196}\right]^{1/2.2} = 6.0 \text{ cm} \qquad (11)$$

3. Calculate the length/diameter ratio (100/6 = 17). Using the dimensionless graph shown in Fig. 5-4, read the volume ratio off the 90% submergence line, as shown by the inset in the figure.

4. Because V_{H_2O}, the volume of water, $= 150$ l, and $0.38 = V_{H_2O}/V_{air}$, then $V_{air} = 150/0.38 = 400$ l min^{-1}.

Solubility of Oxygen

The partial pressure of a gas in the atmosphere is directly proportional to the volume it occupies, and the partial pressure of each gas is the same. In water, the relationship between the volume of a gas in solution and its partial pressure depends on its solubility, or ability to react with water. Oxygen is only moderately soluble, being roughly 28 times less soluble than carbon dioxide and about twice as soluble as nitrogen.

Temperature and salinity are the factors having greatest effect on the solubility of oxygen. Temperature and the solubility of oxygen are related inversely. As the temperature increases, the oxygen-holding capacity of the water decreases, as shown in Table 9-4 (p. 142). Salinity and oxygen solubility also are related inversely. An increase in salinity brings about a decrease in dissolved oxygen. Thus a volume of seawater holds less oxygen than an equal volume of freshwater at the same temperature. This is also summarized in Table 9-4.

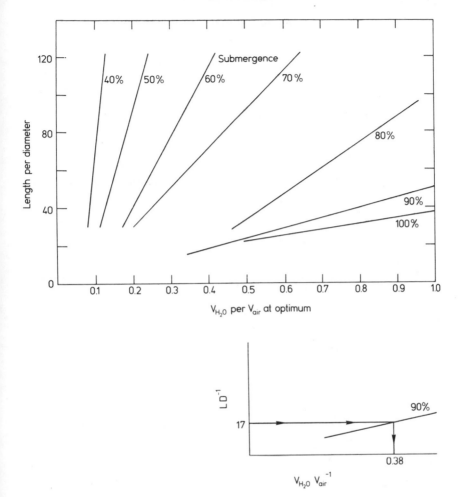

Figure 5-4. Dimensionless graph showing length/diameter ratios of lift pipes versus water flow/air flow volume ratios when air flow is optimum. Drawn from data in Castro et al. (1975). Original.

The flow rate, or rate at which water moves through an aquarium, is controlled by airlift pumping. A constant flow rate of 0.7×10^{-3} m sec^{-1} (1.0 gal ft^{-2} min^{-1}) keeps the oxygen level near saturation at all temperatures and eliminates the possibility of oxygen depletion. An adequate flow rate is especially important in warm-water aquariums because the animals often have higher oxygen requirements. The situation in warm water is complicated further by the lower solubility of oxygen at higher temperatures. It should also be remembered that the filter bed exerts a considerable BOD and competes with the animals for dissolved oxygen.

5.2   RESPIRATION

Water is a difficult medium for respiration. At saturation it contains far less oxygen than air and many times more carbon dioxide. It is a dense, viscous substance and the creatures within its realm must work hard to extract the oxygen they need.

When an organism respires, the oxygen it takes up from the external environment is transported by the vascular system to individual cells within the tissues. At the same time, the vascular system picks up carbon dioxide, an end product of cellular metabolism, and releases it into the environment. Thus *respiration* is a term applicable either to a whole organism or one of its cells. The mechanical process by which respiration is achieved is called *ventilation*. The ventilation rate of aquatic organisms is controlled directly by the concentration of dissolved oxygen in the water (Dejours et al. 1977).

The effectiveness by which oxygen is taken up from water depends on the efficiency of an animal's respiratory pigment. One of the most common respiratory pigments is hemoglobin. The chemical bond between hemoglobin and oxygen can be written in simplified form as

$$Hb + O_2 \rightleftharpoons HbO_2 \qquad (12)$$

At high oxygen concentrations, the hemoglobin (Hb) combines with O_2 to form oxyhemoglobin (HbO_2) and the reaction moves to the right. At low concentrations of O_2 the oxygen molecule is given up and the reaction shifts to the left. If the oxygen concentration is reduced to zero, the hemoglobin relinquishes all of its oxygen. The amount of oxygen that combines with blood at equilibrium at a given partial pressure of oxygen can be plotted as an *oxygen combining curve*. An example of oxygen combining curves for three fishes indigenous to the North Atlantic is presented in Fig. 5-5. The curves show that the blood of an oyster toadfish (*Opsanus tau*) is saturated at lower partial pressures of oxygen than the blood of an Atlantic mackerel (*Scomber scombrus*). Oxygen that is bound up with mackerel blood is released to the tissues readily. Toadfish blood, on the other hand, has a higher oxygen affinity and O_2 is given up to the tissues reluctantly. The oxygen combining curve of the scup (*Stenotomus chrysops*) is intermediate between the toadfish and the mackerel.

Several factors affect the affinity of hemoglobin for oxygen, but two of the most important are carbon dioxide and temperature. Their influence is complicated by diverse physiological requirements and tolerance levels

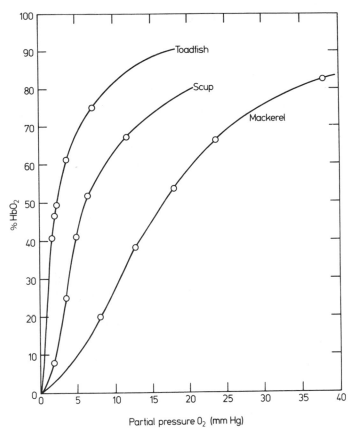

Figure 5-5. Oxygen combining curves for the hemoglobins of three temperate seawater fishes indigenous to the North Atlantic. Redrawn from Hall and McCutcheon (1938).

among species, and also by size and age differences among animals of the same species. The trends outlined in the next sections are, therefore, general rather than specific.

Carbon Dioxide

The amount of free CO_2 in solution is a function of pH, as illustrated in Fig. 7-1 (p. 105). As pH declines, the CO_2 concentration increases. The source of free CO_2 in water is the dissociation of bicarbonate ions. This is discussed further in Chapter 7. The presence of free carbon dioxide in blood influences the shape and position of the oxygen combining curve. General acidity increases as the cells metabolize and release carbon dioxide into the blood. As the blood CO_2 level increases, the affinity of a respiratory pigment for oxygen is reduced, eq. 12 shifts to the left, and oxygen is released to the tissues more readily than it would be otherwise at comparable partial pressures of O_2. As a result, the oxygen combining curve is shifted to the right. This shift is called the *Bohr effect*. Ectotherms show varied responses to the Bohr effect. Many invertebrates show no response at all. Among vertebrates, Lenfant and Johansen (1966) could not find evidence of a Bohr effect in the Pacific dogfish shark *Squalus acanthias* (= *sukleyi*). However, it is known that in many other fishes there is a marked Bohr effect as blood CO_2 increases (Basu 1959, Black et al. 1954, Eddy et al. 1977, Saffran and Gibson 1976).

In the blood of some fishes, there is a dramatic decline in the amount of oxygen that the respiratory pigment can hold at high levels of CO_2. In such cases, the fish's oxygen combining curve is shifted so far to the right that it falls outside the useful range and the the animal suffocates, even in water that is saturated with dissolved oxygen. This phenomenon is called the *Root effect* (Root 1931). An example is shown in Fig. 5-6. The Root effect is simply an exaggerated Bohr effect and many writers do not distinguish between them.

Baines (1975) and Hoar (1975) pointed out that the relationships be-

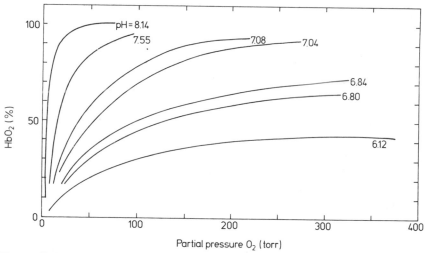

Figure 5-6. Percent saturation of hemoglobin as a function of blood pH in the goosefish (*Lophius americanus*), a seawater species inhabiting the North Atlantic. The drastic shift of the oxygen combining curve to the right as blood pH declines illustrates the Root effect. The curve persists even at high partial pressures of oxygen. Redrawn from Green and Root (1933).

tween free CO_2 and the oxygen-combining properties of blood vary with an animal's ecological situation, and the adjustments that it must make to deliver oxygen to the tissues under varying environmental conditions. Animals with a high Bohr effect have respiratory pigments with a low affinity for oxygen and an oxygen combining curve that is sigmoid-shaped. An example is the curve for the mackerel shown in Fig. 5-5. Animals with blood that has a high affinity for oxygen and a reluctance to release it to the tissues have oxygen combining curves that are rectangular in shape, illustrated by the curve for toadfish blood (Fig. 5-5). The shape of an animal's oxygen combining curve reflects to some degree its life-style, general activity level, and the environment it inhabits. Many schooling fishes like the mackerel move continuously and have high activity levels. Their tissues must receive oxygen readily. This is no handicap because the waters in which they live often contain dissolved oxygen levels that are always near saturation. Toadfish are different from mackerel by being sedentary. They inhabit inshore waters in which the free CO_2 concentration sometimes increases. These increases are often accompanied by low concentrations of dissolved oxygen. At such times, it is to the toadfish's advantage to have blood with a high oxygen affinity.

Temperature

Increased temperature weakens the bond between hemoglobin and oxygen and causes eq. 12 to shift to the left. As a result, the oxygen combining curve shifts to the right because the hemoglobin relinquishes its oxygen more readily. Increased temperature is ordinarily accompanied by increased metabolic rate, and the move to the right has obvious survival value because a higher metabolic rate means greater tissue oxygen demands.

In general, a rise of 10°C in the environment causes the rate of oxygen uptake in an ectothermic animal to double or triple. Kanungo and Prosser (1959) noted that active oxygen consumption in goldfish acclimated to 30°C was 359% higher than in fish acclimated at 10°C. Morris (1962) found that warm acclimation doubled the respiration rate in the cichlid, *Aequidens portalegrensis*. Fry and Hart (1948), Kanungo and Prosser (1959), and Wells (1935) showed that at intermediate temperatures, cold-acclimated fishes have higher respiration rates than warm-acclimated ones. Kanungo and Prosser found that when goldfish acclimated to 10°C were placed in water of 20°C, their standard oxygen consumptions were 26% and their active consumptions 10% higher than fish maintained at 20°C and then acclimated to 30°C.

Thermal tolerance of an ectothermic animal depends on its thermal history, and low dissolved oxygen levels in the environment reduce the tolerance of aquatic animals to temperature changes (Weatherly 1970). Moreover, the level of dissolved oxygen affects the speed at which adaptation to a new temperature takes place. Fry (1947) showed, for example, that thermal acclimation of the brown bullhead (*Ictalurus nebulosus*) occurred within 24 hr when the fish were moved from 20°C to 28°C, provided that adequate dissolved oxygen was available. If the dissolved oxygen level was too low, thermal acclimation was completely inhibited.

Thermal acclimation is thought by most experts to require considerably longer than a few hours. The mechanisms responsible for modifying lethal temperature limits are still unclear, but changes in metabolic enzyme systems play a major role (Somero and Hochachka 1976). Brett (1956), Saunders (1962), Schlieper (1950), and Wells (1935) all demonstrated that thermal acclimation of fishes requires several days. Saunders (1962) showed that complete acclimation of the carp (*Cyprinus carpio*) required 48 hr after transfer from 32°C to 36°C. He considered temperature increases in increments of 1°C 24 hr^{-1} to be a proper acclimation rate. Tyler (1966) used this rate in successfully acclimating two species of freshwater minnows to higher temperatures. When acclimating them to lower temperatures, the rate used was 0.5°C 24 hr^{-1}.

5.3 MANAGEMENT PRACTICES

Good aquarium management practices include maintaining the dissolved oxygen level near saturation at all temperatures and salinities, and making sure that new specimens are acclimated to aquarium temperatures slowly.

Dissolved Oxygen

The flow rate through any aquarium should be a minimum of 0.7 × 10^{-3} m sec^{-1}. This assures dissolved oxygen concentrations near saturation at any temperature or salinity. Dissolved oxygen can be measured by the method outlined in Section 9-5. Temperature and salinity should be determined at the same time and the measured value for dissolved oxygen compared with the value at saturation, as shown in Table 9-4. Increase the flow rate or provide additional aeration with airstones if the measured value is more than 15% below the saturated value. High dissolved oxygen concentrations are especially critical in warm seawater aquariums, where the water holds less oxygen to start with.

Highly active fishes and invertebrates, or species with a known sensitivity to the Bohr effect, should be maintained in aquarium water in which the free CO_2 (determined operationally as pH), temperature, and salinity do not vary. To eliminate the potentially harmful effects of free CO_2, the pH of aquarium seawater should never be allowed to fall below 8.0. Freshwater aquarium animals ordinarily are more tolerant of fluctuations in pH. Nevertheless, the value should not be allowed to go below 7.1. Careful thermal acclimation of new specimens minimizes any Bohr effect caused by temperature. Afterward, the water temperature should not be allowed to vary more than $\pm 1°C$. Sudden increases in temperature of free CO_2 are more harmful than decreases of the same magnitude when the Bohr effect is involved.

Airlift Pump Operation

The mechanism by which air is delivered to the lift pipe is important, particularly in low-lift airlift pumps (Német 1961). If the air is injected through an open-ended tube, the bubbles will be of different diameters. Air bubbles in water rise at different velocities, with larger ones rising more slowly. When bubbles of varied size enter a lift pipe and mix with the water, the result is increased turbulence and reduced performance. More efficient mixing is attained if the air is dispersed into fine bubbles of uniform size, which rise in the lift pipe at the same speed. Moreover, fine bubbles are more efficient gas exchangers because of their larger surface area. A standard carborundum airstone is adequate to reduce turbulence and promote gas exchange in small-diameter lift pipes. In larger airlifts, the air must be dispersed by a perforated plate inserted in the lift pipe directly above the air injection point.

An airlift pump is "pulsatile"; that is, it operates in bursts, even under ideal conditions. As optimum flow is approached, the bursts become shorter and more uniform until they can scarcely be noticed. Air spurting out of a lift pipe with visible pulses ordinarily indicates a surfeit of air, or too small a percent submergence.

Thermal Acclimation

Even short-term deviations from the ambient (acclimation) temperature alter respiratory requirements, produce acid-base imbalances, and cause changes in fluid-electrolyte regulation in aquatic animals (Crawshaw 1977).

Acclimation of newly captured specimens should begin at the ambient temperature in the natural habitat. This prevents stresses caused by temperatures near the lethal limit. In both cold and warm waters, temperatures even slightly higher may cause new animals to exceed their maximum metabolic levels. In temperate climates, ambient temperatures are seasonal and what constitutes acclimation temperature at one time of the year may be lethal 6 months later. For instance, a largemouth bass (*Micropterus salmoides*) caught in winter seldom survives direct immersion in warm water. Such a practice would be less harmful to a bass captured in summer when temperatures in its natural habitat are higher.

Fishes placed suddenly in water of a different temperature undergo thermal shock and exhibit characteristic behavior patterns that are similar in most species. If the new temperature is higher, there is increased activity, loss of equilibrium (including aimless darting about, surfacing, floating in unnatural positions, increased fin movement, and remaining in stationary positions with the tail elevated), and a general increase in ventilation rate, as measured by movements of the opercula (Hoff and Westman 1966). When a fish is subjected to sudden colder temperature, there is loss of equilibrium, increased ventilation rate, and violent convulsions and spasms (Hoff and Westman 1966).

Thermal shock is a major cause of death in newly captured aquarium animals. Tyler (1966) showed that keeping new fishes in their plastic shipping bags and floating the bags in the aquarium even for a short time increased their resistance to thermal stress. The practice is obviously of limited value if there are significant differences between the two temperatures (shipping bag versus aquarium). The temperature in a small bag equilibrates quickly, but the metabolic rate of the animals inside the bag remains unchanged.

The floating technique reduces animal losses when the general condition of the animals is good. If the specimens have been subjected to depleted oxygen and elevated carbon dioxide and ammonia in transit, more harm may be caused by leaving them in the bags and increasing their exposure to these unfavorable conditions.

The safest method of acclimating new animals is to adjust the temperature in the aquarium to the acclimation temperature of the animals prior to collecting or receiving them. If the animals are purchased, a reliable dealer should provide the acclimation temperature beforehand. Once in the aquarium, the animals should be left at the acclimation temperature for at least a week.

Brett (1970) showed that fishes are able to adjust to higher than ambient temperatures faster than when the new temperature is colder than ambient. The proper rate at which to acclimate aquarium animals is 2°C 24 hr^{-1} if the new temperature is higher than ambient, and 1°C 24 hr^{-1} if the new temperature is below ambient. The acclimation period should be accompanied by heavy aeration to assure sufficient dissolved oxygen concentrations.

6

Seawater

Seawater differs from freshwater in the quantity and composition of its dissolved solids, most of which are inorganic salts. The elements known to be present in seawater are shown in Table 6-1.

6.1 DISSOLVED SOLIDS

The sum of the solutes in seawater of normal ionic composition can be determined by measuring the salinity, chlorinity, chlorosity, or density. *Salinity* is traditionally defined as the total amount of solid material dissolved in 1.0 kg of seawater when all carbonate has been converted to oxide, all bromine and iodine replaced by chlorine, and the organic matter completely oxidized. This amount of "solid material" is expressed in grams, and salinity is measured in g kg^{-1}, or parts per thousand (ppt). The symbol for salinity is S; ‰ stands for ppt. Salinity includes both the inorganic ions in solution and the organic compounds. Thus a reading of 35 S ‰, the normal value for seawater, is an expression of all dissolved solids—ions such as sodium and chloride, and also organic phosphorus and nitrogen, plant pigments, vitamins, and so forth.

Salinity is difficult to measure by direct chemical methods, as its definition implies, so it is usually defined in terms of another variable, chlorinity (Cl ‰). The relationship, as demonstrated by Wooster et al. (1969), is

$$S\text{ ‰} = 1.80655 \ (Cl\text{ ‰}) \qquad (13)$$

Chlorinity is a measure of the halogen concentration in a sample of seawater, and it is easier to determine analytically than salinity. The usual method is by titrating a sample of seawater with silver nitrate, using potassium dichromate as an indicator. By modern definition, *chlorinity* is the numerical value in grams per kilogram of a seawater sample identical to the number giving the mass in grams of atomic weight silver just necessary to precipitate the halogens in 0.3285233 kg of the sample. In the past, oceanographers considered chlorinity to be the total amount of chlorine, bromine, and iodine in grams contained in 1.0 kg of seawater, assuming

89

Table 6-1. Elemental Composition of Seawater. Adapted from Bowen (1966)

Element	Chemical Form	Concentration (mg l^{-1})
Ag	$AgCl_2^-$	0.0003
Al		0.01
Ar	Ar	0.6
As	AsO_4H^{2-}	0.003
Au	$AuCl_4^-$	0.000011
B	$B(OH)_3$	4.6
Ba	Ba^{2+}	0.03
Be		0.0000006
Bi		0.000017
Br	Br^-	65
C	CO_3H^-, organic C	28
Ca	Ca^{2+}	400
Cd	Cd^{2+}	0.00011
Ce		0.0004
Cl	Cl^-	19,000
Co	Co^{2+}	0.00027
Cr		0.00005
Cs	Cs^+	0.0005
Cu	Cu^{2+}	0.003
F	F	1.3
Fe	$Fe(OH)_3$	0.01
Ga		0.00003
Ge	$Ge(OH)_4$	0.00007
H	H_2O	108,000
He	He	0.0000069
Hf		<0.000008
Hg	$HgCl_4^{2-}$	0.00003
I	I^-, IO_3^-?	0.06
In		≪0.02
K	K^+	380
Kr	Kr	0.0025
La		0.000012
Li	Li^+	0.18
Mg	Mg^{2+}	1350
Mn	Mn^{2+}	0.002
Mo	MoO_4^{2-}	0.01
N	Organic N, NO_3^-, NH_4^+	0.5
Na	Na^+	10,500
Nb		0.00001
Ne	Ne	0.00014

Table 6-1.　(Continued)

Element	Chemical Form	Concentration (mg l^{-1})
Ni	Ni^{2+}	0.0054
O	H$_2$O, O$_2$, SO$_4$$^{2-}$	857,000
P	PO$_4$H^{2-}	0.07
Pa		2×10^{-9}
Pb	Pb^{2+}	0.00003
Ra		6×10^{-11}
Rb	Rb$^+$	0.12
Rn	Rn	6×10^{-16}
S	SO$_4$$^{2-}$	885
Sb		0.00033
Sc		<0.000004
Se		0.00009
Si	Si(OH)$_4$	3
Sn		0.003
Sr	Sr^{9+}	8.1
Ta		<0.0000025
Th		0.00005
Ti		0.001
Tl	Tl$^+$	<0.00001
U	UO$_2$(CO$_3$)$_3$$^{4-}$	0.003
V	VO$_5$H$_3$$^{2-}$	0.002
W	WO$_4$$^{2-}$	0.0001
Xe	Xe	0.000052
Y		0.0003
Zn	Zn^{2+}	0.01
Zr		0.000022

that the bromine and iodine had been replaced by chlorine. The essential features of this definition still apply when defining chlorosity (see below). It is ordinarily not necessary to determine chlorinity in routine aquarium maintenance.

The halogen concentration of seawater is sometimes expressed as chlorosity, particularly by European chemists. Chlorosity means the same thing as chlorinity, but is given in grams per litre instead of grams per kilogram. Because volumetric measurements are affected by temperature, *chlorosity* is defined as the total amount of chlorine, bromine, and iodine in grams contained in 1.0 l of seawater *at 20°C*, assuming that the bromine and iodine have been replaced by chlorine. Chlorosity readings will be 2%–3%

greater than chlorinity determinations made on the same sample because a litre of seawater has a greater mass than a litre of freshwater.

The terms density and specific gravity often are used interchangeably by aquarists, but they are not synonyms. The *density* of seawater refers to its mass per unit volume and is expressed as grams per ml or cubic centimetre. Specific gravity is the ratio of a given volume of seawater and an equal volume of distilled water at 4°C. Specific gravity, being a ratio, has no units; it is dimensionless. The distilled water must be specified as being at 4°C because pure water reaches maximum density at this temperature.

The density of seawater varies with changes in temperature and salinity. Pressure, a third factor, is inconsequential in aquarium operation. Even in oceanographic work, pressure is of minor importance because seawater is nearly incompressible. Salinity affects density because the mass of water increases when ions are added to it. Increasing the salinity by 1 ‰, for example increases the density by about 0.8 ‰.

Temperature has a more pronounced effect on density than does salinity. The density of full-strength seawater ($S = 25$ ‰ or more) decreases with increasing temperature, as illustrated in Fig. 6-1. In other words, cold seawater is more dense than warm seawater of the same salinity (salinity, being a weight-per-weight relationship, functions independently of temperature). As seawater is made brackish by dilution with freshwater, it

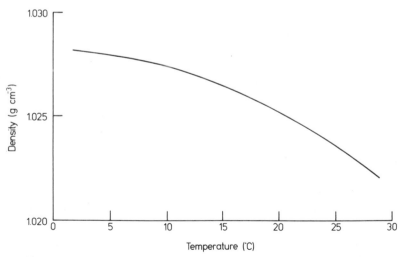

Figure 6-1. Density of seawater versus temperature. Redrawn from Anikouchine and Sternberg (1973).

starts to behave more like pure water and becomes less dense if cooled below the temperature at which it reached maximum density.

Density is not an easy concept to understand, nor is its measurement simple and direct. To make accurate density determinations, the effects of temperature and salinity must be considered both independently and in combination (Williams 1962). Salinity is the most reliable means of measuring the dissolved solid concentration in seawater. Density and specific gravity measurements (the latter requiring temperature correction) are less reliable.

The normal chlorinity of seawater is considered to be 19 ‰, which equals a salinity of 34.325 ‰. The normal specific gravity of seawater is considered to be 1.024 at a salinity of 34 ‰ and a temperature of 20°C.

6.2   FUNCTIONS AND UPTAKE OF ELEMENTS

Organisms living in brackish water and seawater aquariums are affected by the ionic strength and the composition of the medium. Many of the elements present in solution are necessary to carry out vital biochemical processes. Bowen (1966) listed the main cellular functions of elements as electrochemical, catalytic, and structural. Elements function electrochemically when they serve as metabolic energy sources. It is likely that all essential elements function as enzyme activators and help regulate the rates of biochemical reactions. In this respect, they demonstrate catalytic functions. Many elements are necessary in the synthesis of such substances as proteins and amino acids. Here the function is structural and the element is a necessary constituent in the final product.

Most, if not all, of the known elements are found in natural waters. Many have no measurable effects and probably are not essential. Arnon and Stout (1939) listed three factors that determine whether a given element X is essential: (1) The organism cannot grow or complete its life cycle if X is not available; (2) X cannot be completely replaced by another element; and (3) X directly influences the metabolic functions of the organism.

Elements enter animals from the environment by two mechanisms: (1) diffusion, or passive transport; and (2) active transport. *Passive transport*, in which an element moves from a greater concentration in the water into the more dilute intracellular fluid, needs no explanation. *Active transport* is the selective extraction of elements from the water, often against a concentration gradient. It is closely correlated with temperature and a rise of 10°C increases the rate of active transport by about 100% (Bowen 1966).

The amount of dissolved oxygen present is also a factor: when respiration is inhibited, so is the active transport of ions from the medium.

The pros and cons of adding trace elements to closed-system aquariums have been debated for years. Perhaps there is no single answer, although a few general comments may help to clarify the situation. Ordinarily, an aquarium is an elemental sink: N, P, S, Ca, Mg, Na, and many other elements increase with time. They originate from the excretory products of animals and the exudates of algae. The numbers of bacteria also increase with time, before eventually attaining equilibrium. As bacterial cells die and lyse, still more elements are added to the medium. The real question is not whether substances become depleted with time, but whether they remain in biologically useful form when returned to the water by these pathways.

Some elements are depleted if physical adsorption methods (activated carbon and foam fractionation) are used, or if plants are cultured in the water. Still others are oxidized by ozonation, and perhaps even by UV irradiation. Thus the composition of a captive seawater medium changes with time.

Animals probably suffer less from such changes than plants. Animals obtain many of their nutrients secondhand by eating plants and other animals. Plants must obtain nutrients directly from the water, and their growth is governed by the continued presence of trace elements in useful states. According to O'Kelley (1974), the following elements in inorganic form (plus C, H, and O) are essential to at least one species of freshwater or marine algae: N, P, K, Mg, Ca, S, Fe, Cu, Mn, Zn, Mo, Na, Co, V, Si, Cl, B, and I. Of these, N, P, Mg, Fe, Cu, Mn, Zn, and Mo are required by all algae and are not replaceable even in part by other elements. On the other hand, Sr, K, and Ca can be replaced in certain circumstances by another ion of like chemical charge (e.g., Sr for Ca).

It can be argued that all of these substances probably are present in sufficient amounts as contaminants in the other salts. However, there is no guarantee that critical elements will always be available in sufficient amounts to promote algal growth. Lush growths of algae are more apt to occur if trace elements are added routinely along with the other salts during each biweekly 10% partial water change. This is easily accomplished by including trace elements in artificial seawater formulations.

6.3   ARTIFICIAL SEAWATER

Prepackaged mixes for small volumes of artificial seawater can be purchased in tropical fish stores (Fig. 6-2). Large volumes must be prepared

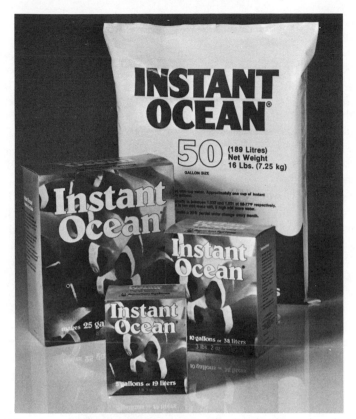

Figure 6.2. A prepackaged artificial seawater mix. Courtesy Aquarium Systems, Inc.

on site from published formulas. Prepackaged mixes or formulas that omit elements essential to algae are not recommended, unless the culture of algae is undesirable or unnecessary. The brand or formula selected should allow for the major salts to be present in ratios and concentrations approximating natural seawater. Other considerations are secondary.

Mixing Small Volumes

The most useful prepackaged mixes can be hydrated in a single step using tap water. Some need to be aerated for several hours before the pH equilibrates. Others, such as Instant Ocean® Synthetic Sea Salts,* come rapidly to equilibrium, allowing animals and plants to be added immedi-

*Aquarium Systems, Inc. Twinbrook, Mentor, OH 44060

ately. When using most brands, check the pH hourly and do not put living organisms into the water until the pH has equilibrated at 8.1–8.3.

Of the prepackaged mixes, Instant Ocean® has proved satisfactory for culturing a great number of animal and plant species, both in the laboratory and in public aquariums. This medium also is used routinely in bioassay work. Other brands may be suitable too, but none has been tested so thoroughly as Instant Ocean® under the rigidly controlled conditions that are required in marine research.

Follow the instructions carefully when using any of the prepackaged mixes. Artificial sea salts can be hydrated in clean containers made from inert materials. Spare aquarium tanks, polyethylene cannisters or jerry cans, or plastic garbage cans with tight-fitting lids are adequate. The finished solution should be covered to prevent concentration of the salts from evaporation, and contamination by dust and airborne toxicants. Aquarium Systems, Inc. manufactures a special polypropylene mixing container holding about 400 l (Fig. 6-3). The device is equipped with a pump for transferring the finished solution directly to aquarium tanks.

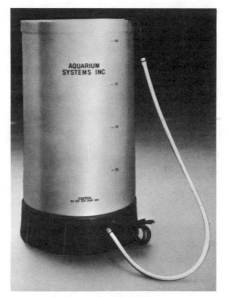

Figure 6-3. A mixing tank for 400 l of artificial seawater. The unit is equipped with a pump for transferring the finished solution to aquarium tanks. Courtesy Aquarium Systems, Inc.

Mixing Large Volumes

Large volumes of artificial seawater should be mixed in stages over a three-day period instead of in a single step. A good procedure is to dissolve the major salts one at a time. Dissolution is quicker if the container holding the salts is flooded continuously with tap water at 30°C. Minor salts can be added on the second day and trace salts on the third day. Minor and trace components often precipitate if added before the major salts have been diluted to volume. Heavy aeration can be used to disperse all the elements evenly throughout the solution, and by the fourth day the medium should be suitable for culturing live marine organisms. The formula given here is a modified version of an older one published by Segedi and Kelley (1964). It has proved to be reliable and easily reproducible.

Mixing Equipment

A well designed mixing arrangement for artificial seawater consists of a dissolving chamber for hydrating salts, and a storage vat underneath to hold the finished solution (Fig. 6-4). The dissolving chamber should be equipped with hot and cold tap water discharging through a common valve. An in-line thermometer placed after the mixing valve is useful for regulating water temperature. An electric mixer (not shown in the figure) mounted on the side of the chamber aids in dissolving the salts. The storage vat should contain a large airlift for circulating the water. If an airlift is not used, a mechanical pump is required. A suitable airlift can be made from a straight length of PVC pipe (10 cm in diameter or larger) with a hole drilled near one end. The size of the hole should be just large enough to make a tight fit for the airline (0.95 cm diameter flexible plastic tubing works well). No dispersion device is necessary because the purpose of the airlift is simply to circulate water through the vat. The airlift can be fixed permanently in place over the lowest part of the vat where salts accumulate, or anchored to the bottom with lead weights. It is not necessary for the top of the airlift to be above the surface of the water when the vat is full.

In large installations, it is convenient to add sodium chloride, the major constituent in seawater, as a concentrated brine instead of in crystalline form. This can be done with automatic dissolving equipment developed and sold by the International Salt Company, Inc.* (Fig. 6-5). Bagged salt

---

*International Salt Co., Clarks Summit, Pa. 18411

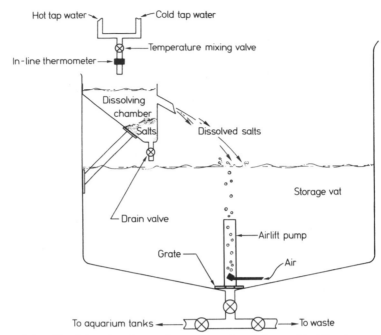

Figure 6-4. Dissolving chamber and storage vat for mixing and holding large volumes of artificial seawater. Original.

is awkward to handle in large quantities, and requires considerable storage space. Bulk (unbagged) salt is easy to move around as a liquid and requires less storage space, provided that the necessary equipment is incorporated into the plant design. Bulk salt is also cheaper to purchase. Concentrated brine can be pumped easily to artificial seawater storage vats, and its addition in this form is quicker and more accurate than the procedure of emptying large amounts of bagged NaCl into the dissolving chamber.

A diagrammatic illustration of a Sterling Brinomat® is shown in Fig. 6-5. Bulk salt is stored in the hopper. The hopper is refilled pneumatically through a connecting stainless steel pipe leading outside the building to a truck loading ramp. Tap water entering the Brinomat® flows down through the lower portion of the salt bed. As the water moves downward, it forms a brine solution of increasing strength. Just above the bottom it reaches full saturation, provided the hopper is kept full. Dissolved salt is replaced automatically by dry salt from the hopper. It is important to note that the brine becomes saturated *before* it reaches the bottom of the Brinomat®. The lower portion of the salt bed never dissolves and acts as a mechanical filter which removes insoluble impurities from the effluent

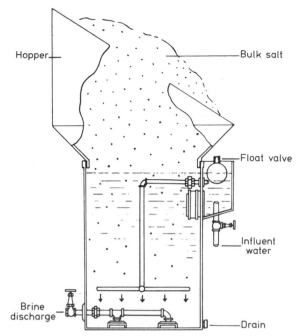

Figure 6-5.   The Sterling Brinomat®. Courtesy International Salt Co., Inc.

brine. As brine is drawn from the Brinomat® and transferred to the mixing vat, a float valve opens the tap water line and brine making continues automatically. In seawater culture, a Brinomat® must be equipped with an accurate flow meter to measure the amount of solution passing into the storage vat. Brass meters are preferable, because they are corrosion resistant. The minute amount of copper that may leach from a brass meter is insignificant from a toxicity standpoint. Besides, the tap water used to hydrate the dry salt will have passed through many metres of copper pipe before it reaches the storage vat. All-brass, plastic, or stainless steel pumps are recommended for pumping brine or seawater.

The quantities of major salts required for 30,240 l (8000 U. S. gal) of artificial seawater at a salinity of 35 ‰ (specific gravity of about 1.025 at 20°C) are shown in Table 6-2. Sodium chloride, if added from a Brinomat®, is normally at saturation and the correct amount is 2669 l (706 U. S. gal). However, the specific gravity of the Brinomat® effluent will vary if the hopper is not kept full. Major salts should always be purchased in 45 kg (100 lb) plastic-lined paper bags. Minor and trace salts should be purchased and stored in tightly sealed glass or polypropylene bottles.

Table 6-2.   Major Salts for 30,240 l of Artificial Seawater. Modified from Segedi and Kelley (1964)

Salt*	Amount (kg)
Sodium chloride (NaCl)	834.60
Magnesium sulfate ($MgSO_4 \cdot 7H_2O$)	208.65
Magnesium chloride ($MgCl_2 \cdot 6H_2O$)	163.29
Calcium chloride ($CaCl_2 \cdot 2H_2O$)	41.73
Potassium chloride (KCl)	18.14
Sodium bicarbonate ($NaHCO_3$)	6.35

* Use technical grade salts.

Procedure for Mixing the Major Salts

1. Be sure the storage vat is clean. If it is dirty, scrub the walls and floor with a stiff bristle brush and hose it out with tap water. Do not use prepared cleansers. Stained areas can be cleaned with a strong solution of sodium bicarbonate and warm water, then rinsed with tap water.

2. Turn on the airlift pump and close the drain valve from the vat.

3. Turn on the tap water to the dissolving chamber and adjust the temperature to 30°C. Wait until there are about 30 cm of water in the bottom of the vat before proceeding.

4. Check the specific gravity or salinity of brine from the Brinomat® effluent and determine the amount needed from Table 6-3.

5. Weigh the correct amounts of the other major salts shown in Table 6-2. Fractions of bags should be assembled in clean plastic containers.

6. Turn on the Brinomat® pump and add the correct amount of brine.

7. Add the rest of the salts individually to the dissolving chamber while the tap water is running. Wait until each batch dissolves before adding the next. Use this procedure:

    (*a*) Fill the dissolving chamber half-full with a component salt after turning on the electric mixer.

    (*b*) Refill the chamber half-full and repeat *a* until all the salts are dissolved.

    (*c*) Count the empty bags and containers as a check to be sure that no salts have been forgotten.

8. Continue filling the storage vat with 30°C water until the salinity reaches 34 ‰. (Once the correct level in the vat has been found the first time, the water level should be marked permanently with epoxy paint.)

**Table 6-3. Volume of Brine at Different
Specific Gravity Values Needed to Make
30,240 l of Artificial Seawater.***

Specific Gravity	Volume (l)	Volume (U. S. gal)
1.151	3655	967
1.154	3599	952
1.156	3546	938
1.158	3497	925
1.160	3447	912
1.162	3398	899
1.164	3349	886
1.167	3304	874
1.169	3258	862
1.171	3209	849
1.173	3168	838
1.175	3126	827
1.177	3084	816
1.180	3047	806
1.182	3005	795
1.184	2967	785
1.186	2930	775
1.188	2895	766
1.190	2858	756
1.193	2824	747
1.195	2790	738
1.197	2756	729
1.199	2722	720
1.202	2688	711
1.203 99% saturated	2669	706

* Calculations based on the properties of brine at 15.56°C.

Procedure for Mixing the Minor Salts

1. One day after the major salts have been mixed, weigh the proper amount of each minor salt (Table 6-4). Combine the salts in a large beaker. *Do not* add water.

2. Sprinkle the dry mixture on the surface of the water in the storage vat.

3. Use the minor salts immediately. If they remain in the beaker longer

**Table 6-4.   Minor Salts for 30,240 l of Artificial
Seawater. Modified from Segedi and Kelley
(1964)**

Salt*	Amount (g)
Strontium chloride ($SrCl_2 \cdot 6H_2O$)	600
Manganese sulfate ($MnSO_4 \cdot H_2O$)	120
Sodium phosphate ($NaH_2PO_4 \cdot 7H_2O$)	120
Lithium chloride (LiCl)	30
Sodium molybdate ($Na_2MoO_4 \cdot 2H_2O$)	30
Sodium thiosulfate ($Na_2S_2O_3 \cdot 5H_2O$)	30

* Use reagent grade salts.

than 2 hr, noticeable reactions that alter the chemical states of the components will take place.

Procedure for Mixing the Trace Salts

1. Fill a clean 19-l (5.0 U. S. gal) polypropylene bottle with about 2.0 l of distilled water.
2. Weigh the trace salts (Table 6-5) and place them in individual, labeled beakers. Add sufficient distilled water to dissolve.
3. Add each solution to the bottle. Rinse the beakers with distilled water and add this water also.
4. Add distilled water to make about 12 l. Place an airstone in the bottle and aerate the solution moderately until it is needed. Add the trace element solution to the storage vat one day after the minor salts have been added.

**Table 6-5.   Trace Salts for 30,240 l of Artificial
Seawater. Modified from Segedi and Kelley
(1964)**

Salt*	Amount (g)
Potassium bromide (KBr)	812.7
Aluminum sulfate ($Al_2[SO_4]_3 \cdot 18H_2O$)†	26.0
Rubidium chloride (RbCl)	4.5
Zinc sulfate ($ZnSO_4 \cdot 7H_2O$)	2.9
Cobalt sulfate ($CoSO_4 \cdot 7H_2O$)	2.7
Potassium iodide (KI)	2.7
Cupric sulfate ($CuSO_4 \cdot 5H_2O$)	0.3

* Use reagent grade salts.
† Heat required to dissolve.

6.4 MANAGEMENT PRACTICES

Major salts used to make artificial seawater should be stored in a room with low humidity. Salts left over after mixing the major components of the modified Segedi-Kelley formula are best kept in labeled polypropylene containers with tight-fitting lids, rather than left lying about in opened bags. Minor and trace salts should be stored in polypropylene or glass containers with tight-fitting lids. It is a good practice not to order more material than can be used in 6 months. Scales and analytical balances used to weigh salts should be recalibrated frequently.

CHAPTER 7

Buffering

The chemical interactions of free carbon dioxide, water, and mineral carbonates constitute the carbon dioxide system in natural waters. Mineral carbonates buffer the water by preventing extreme fluctuations in pH, which may be detrimental to animals, plants, and filter bed bacteria. The specific effects that changes in pH have on the physiology of aquatic organisms are poorly understood, but in fishes low pH values are known to interfere with oxygen uptake from water, as shown in Chapter 5.

Terms commonly associated with the carbon dioxide system are buffer, alkalinity, hardness, and pH. A *buffer* is any substance that inhibits a change in hydrogen ion (H^+) concentration in the water. The principal buffers are the ions carbonate (CO_3^{2-}) and bicarbonate (HCO_3^-). They neutralize any addition or withdrawal of free carbon dioxide (CO_2) and maintain constant pH levels by repressing fluctuations in the hydrogen concentration. Bicarbonate ion is the dominant constituent in the carbon dioxide system throughout the natural pH range of aquarium water (7.1–7.8 in freshwater, and 8.0–8.3 in brackish water and seawater), as shown in Fig. 7-1.

Alkalinity is the sum of negative ions reacting to neutralize hydrogen ions when an acid is added to water. Carbonate and bicarbonate ions are the most important, although boric acid accounts for about 10% of the buffer capacity of seawater (Skirrow 1975). Alkalinity is normally expressed as meq l^{-1}. Alkalinity values in seawater range from 2.1 to 2.5 meq l^{-1}. The alkalinity of freshwater is lower and more variable because it is less strongly buffered.

Hardness is a term limited to freshwater. It gives a measure of the total concentration of calcium and magnesium and, like alkalinity, is expressed in units of meq l^{-1}. Other cations in freshwater have a negligible effect: sodium and potassium do not contribute to hardness because of their high solubilities, and other species are present only in trace amounts. Freshwaters containing low concentrations of calcium and magnesium are classified as *soft*.

The *pH* value is a measure of the hydrogen ion concentration, resulting from changes in alkalinity. It is defined as $1/\alpha H^+$. Water of pH less than 7.0 at 25°C is acidic; greater than this, basic. A pH of exactly 7.0 is neutral

104

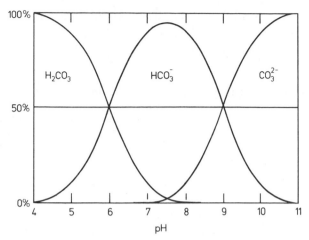

Figure 7-1. The relative abundance of H_2CO_3, HCO_3^-, and CO_3^{2-} in seawater as a function of pH. Redrawn from Weyl (1970).

(neither acidic nor basic). When one of the dissociation products of a reaction is H^+, the solution is acidic and the pH decreases. When a chemical reaction produces OH^- ions, it is basic and there is a rise in pH.

7.1 CARBONATE AND BICARBONATE IONS

Carbonate and bicarbonate ions in aquarium water are derived from three sources: (*1*) the reaction of free CO_2 with water; (*2*) the reaction of mineral carbonates with free CO_2 and water; and (*3*) bacteriological reduction processes.

Reaction of Free CO_2 with Water

Free CO_2 is very soluble. It enters aquarium water from the atmosphere at the air-water interface and is naturally present in solution as a by-product of metabolism. Free CO_2 reacts with water to produce carbonic acid (eq. 14). Carbonic acid then dissociates to release hydrogen and bicarbonate ions (eq. 15). Bicarbonate ions can dissociate to give more free hydrogen, in addition to carbonate ions (eq. 16). The composite reaction is extremely pH sensitive and shifts to the right as pH increases:

$$\overset{\text{(14)}}{} \qquad \overset{\text{(15)}}{}$$

$$CO_2(g) \rightleftharpoons CO_{2(aq)} + H_2O \rightleftharpoons H_2CO_3 \rightleftharpoons H^+ \qquad (14\text{--}16)$$

$$\overset{\text{(16)}}{} $$

$$+ HCO_3^- \rightleftharpoons H^+ + CO_3^{2-}$$

In hard freshwaters and seawater, both of which are well-buffered, the carbonic acid-bicarbonate equilibrium (eq. 15) is dominant. This is clear from the preponderance of bicarbonate ions within the pH range 7.1–8.3, as shown in Fig. 7-1.

Reaction of Mineral Carbonates with Free CO_2 and Water

The second source of carbonate and bicarbonate ions in aquarium water is the reaction of mineral carbonates with free CO_2 and water. In hard freshwaters and seawater, much of the carbonate material that potentially affects pH is bound up with calcium and magnesium. These mineral carbonates act as reserves of potential bicarbonate ions, ready to dissociate and neutralize any increase in hydrogen ions.

As aquarium water gradually becomes more acidic from biological oxidation processes, mineral carbonates are brought into solution by reaction with free CO_2 and water. In the case of calcium carbonate, the dissociation products are free calcium and bicarbonate ions (eq. 17).

$$CaCO_3 + CO_2 + H_2O \rightleftharpoons Ca^{2+} + 2HCO_3^- \qquad (17)$$

The dissolution of carbonate minerals in seawater is, by itself, inadequate to sustain the pH within the optimum range of 8.0–8.3. Solutions to the problem are discussed in Section 7.4.

Bacteriological Reduction Processes

Carbonate and bicarbonate ions are formed as the result of bacteriological reduction processes—in aquarium water, especially during dissimilation of inorganic nitrogen by anaerobic bacteria (Section 1.3). If water circulation is poor, there is localized alkalinization at reaction sites on detritus particles and a rise in pH. This can be expected at the surfaces of filter beds with a heavy detritus accumulation and low flow rates. If the pH value at the surface of the filter bed approaches 9.0, ammonia produced by heterotrophic bacteria reacts with calcium ions and produces a precipitate of calcium carbonate plus ammonia. According to Berner (1968), the reaction in nature is

$$NH_3 + Ca^{2+} + HCO_3^- \rightarrow CaCO_3 + NH_4^+ \qquad (18)$$

If the pH is within the normal range, free CO_2 generated by the combined respiration of animals and aerobic microorganisms reacts with water and ammonia to produce carbonate and bicarbonate ions. The reactions in nature, again from Berner (1968), are as follows:

$$CO_2 + NH_3 + H_2O \rightarrow NH_4^+ + HCO_3^- \qquad (19)$$

$$CO_2 + 2NH_3 + H_2O \rightarrow 2NH_4^+ + CO_3^{2-} \qquad (20)$$

7.2   SOLUBILITY OF MINERAL CARBONATES

Buffer activity is partly a function of the solubility of calcium and magnesium carbonates; solubility, in turn, depends on the degree of saturation in the water. The solubility of mineral carbonates is affected adversely by (*1*) the presence of magnesium in solution; and (*2*) the presence of dissolved organic carbon (DOC).

Magnesium

There are two mechanisms in seawater by which magnesium inhibits the dissolution and precipitation of carbonate minerals. The first is by acting as a surface poison at the water-crystal interface; the second is through formation of magnesian calcite overgrowths on the crystals, which alter the surface solubility. Folk (1974) and Lippmann (1960) suggested that magnesium inhibited crystal growth on calcite surfaces by acting as a surface poison and by being adsorbed as hydrated ions at growth sites on the crystal surfaces. In addition, magnesium may be incorporated into a growing crystal to such a degree that its solubility is increased. The new surface of the crystal, being now composed of magnesian calcite and not pure calcite, would equilibrate with the magnesium ions in seawater. The overgrowth that results would cause the surface of the crystal to become destabilized and therefore more soluble (Berner 1975, Plummer and Mackenzie 1974, Weyl 1967).

Chave et al. (1962) found that high magnesian calcite had the greatest solubility of the mineral carbonates, followed in order of decreasing solubility by aragonite, low magnesian calcite, and pure calcite. Dolomite, ordinarily consisting of calcium and magnesium in a 1:1 ratio, was the least soluble.

In practical terms, formation of magnesian calcite overgrowths evens out the solubilities of the different carbonate minerals and lessens the importance of their original compositions. As Berner (1975) and Weyl (1967) noted, the overgrowth reaches equilibrium with the concentration of magnesium in the surrounding seawater. To aquarists, this means that the water ultimately buffers the surfaces of the minerals and not the other way around. Thus the original composition of a mineral carbonate plays a limited role in terms of its usefulness in maintaining the alkalinity of seawater at normal values. The presence of dissolved organic carbon only compounds the problem.

Dissolved Organic Carbon

Barcelona et al. (1976), Chave (1965), and Chave and Suess (1967, 1970) demonstrated that naturally occurring DOC inhibited the precipitation of calcium carbonate from supersaturated natural and artificial seawaters. Meyers and Quinn (1971) and Suess (1970) described the adsorption of fatty acids and natural lipids onto calcite surfaces. As DOC coated the surfaces of carbonate particles, it reduced ionic exchange sites.

7.3 DECLINE IN pH

Respiration by animals and plants produces free CO_2. Some bacteriological processes are oxidative and produce acids, whereas others, like denitrification, result in a rise in pH. Plant photosynthesis is also reductive, causing pH increases. A summary of common biological processes that affect pH is presented in Table 7-1.

Biological oxidation in aquariums exceeds reduction overall, and there is a gradual decline in alkalinity and pH. Mineralization of organic carbon compounds and nitrification account for the bulk of acid-forming processes of bacteriological origin. Gundersen and Mountain (1973) showed how accumulating nitrate affected the alkalinity of surface seawater. If the reactions for nitrification are written to illustrate a gain or loss of electrons, the oxidation of ammonium ion is shown by

$$NH_4^+ + OH^- + 2H_2O \rightarrow H^+ + NO_2^- + H_2O + 6[H^+ + e^-] \quad (21)$$

and the oxidation nitrite by

$$NO_2^- + H^+ + H_2O \rightarrow H^+ + NO_3^- + 2[H^+ + e^-] \quad (22)$$

The overall reaction is

$$NH_4^+ + OH^- + 3H_2O \rightarrow H^+ + NO_3^- + H_2O + 8[H^+ + e^-] \quad (23)$$

Table 7-1. Biologically Mediated Reactions Affecting pH in Natural Water Systems. After Weber and Stumm (1963)

Process	Reaction	Effect on pH
Photosynthesis	$6\,CO_2 + 6\,H_2O \rightarrow C_6H_{12}O_6 + 6O_2$	increase
Respiration	$C_6H_{12}O_6 + 6O_2 \rightarrow 6\,CO_2 + 6\,H_2O$	decrease
Methane fermentation	$C_6H_{12}O_6 + 3\,CO_2 \rightarrow 3\,CH_4 + 6\,CO_2$	decrease
Nitrification	$NH_4^+ + 2\,O_2 \rightarrow NO_3^- + H_2O + 2\,H^+$	decrease
Denitrification	$5\,C_6H_{12}O_6 + 24\,NO_3 + 24\,H^+ \rightarrow 3\,0$ $CO_2 + 12\,N_2 + 42\,H_2O$	increase
Sulfide oxidation	$HS^- + 2\,O_2 \rightarrow SO_4^{2-} + H^+$	decrease
Sulfate reduction	$C_6H_{12}O_6 + 3\,SO_4^{2-} + 3\,H^+ \rightarrow 6\,CO_2$ $+ HS^- + 6\,H_2O$	increase

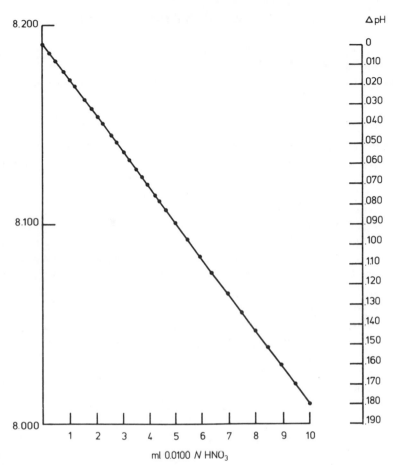

Figure 7-2. Change in pH of open ocean surface water titrated with 0.01 N HNO_3. Redrawn from Gundersen and Mountain (1973).

Equation 23 demonstrates that the conversion of 1 mole of ammonium ion results in the formation of 1 mole of nitrate ion and 1 mole of hydrogen ion. In other words, the end product of nitrification is, strictly speaking, *nitric acid*, rather than nitrate ion, and its effect is to reduce the buffer capacity of the water. Figure 7-2 illustrates how nitrate ion formation is accompanied by an equivalent amount of hydrogen ions. Seawater titrated with 0.01 N HNO_3 with simultaneous pH monitoring produces a linear curve, indicating a drop in pH as nitrification proceeds.

7.4 MANAGEMENT PRACTICES

The effects that low pH values have on animals are poorly understood. Most studies reported in the literature have dealt with pH levels considerably below normal, and the results are not applicable to aquarium keeping. Elevated concentrations of free CO_2, which are accompanied by depressed pH values, are known to affect the affinity of fish blood for hemoglobin (Section 5.2), but most of the experimental work has been performed using waters that were acidic or nearly neutral. To repeat, safe pH ranges are 7.1–7.8 for freshwater, and 8.0–8.3 for brackish water and seawater.

The pH of freshwater aquariums can be maintained above neutral by addition of calcareous gravel to the filter bed. The ratio of silica gravel to calcareous material should be 4:1. When excess detritus is removed during partial water changes, and activated carbon is used occasionally to adsorb some of the DOC from solution, oxidation processes by bacteria are reduced and the chances of the pH falling below 7.1 are slight. It is a common practice for freshwater hobbyists to maintain tropical fishes in acidic water on the assumption that it duplicates the natural environment (many rain forest rivers and streams are acidic). It should be remembered that captive animals are subjected to a number of physiological stresses to start with, and acidic water presents another by increasing the partial pressure of CO_2 in solution.

The maintenance of correct pH levels in seawater or brackish water is more complex. According to Hirayama (1970), when calcareous gravels are relied on exclusively to buffer water in closed-system seawater aquariums, the water eventually equilibrates at about pH 7.5 and alkalinity 1.0 meq l^{-1}. Both values are below recommended levels. Partial water changes at a rate of 10% every 2 weeks, and regular addition of sodium carbonate (Na_2CO_3) or sodium bicarbonate ($NaHCO_3$) are usually necessary to keep the pH and alkalinity values within normal ranges. The use of powdered limestones is not recommended. Calcium in the lime may change the ratio of major cations when used for long periods of time, as noted by Breder and Smith (1932). Cooper (1932) reported that maintenance of pH at the Plymouth Aquarium in England by addition of lime eventually increased the calcium level to 0.62 g l^{-1}, compared with 0.39 g l^{-1} in seawater off Plymouth.

Sodium, as the dominant cation in seawater, can be added as carbonate or bicarbonate salts for longer periods before the cation balance is upset. Breder and Smith (1932) estimated that if all the bicarbonate ion in seawater at the New York Aquarium were to be replaced entirely over a 2.5-yr period using sodium bicarbonate, the sodium level would have increased by a mere 0.5%. If lime were used instead, the calcium level would have increased by 10%, or by a factor of 20 above the molar concentration of calcium in fresh seawater. As Breder and Smith pointed out, the dangers associated with use of lime are not immediate, but it is best to keep the possibility remote.

8

Toxicity and Disease Prevention

Most of the toxicity problems encountered in aquariums can be traced to nitrogenous compounds. Ammonia is the most toxic, followed by nitrite. Nitrate is the least toxic, although its presence in high concentrations for prolonged periods may have deleterious effects that have been overlooked. Nitrite appears to be far less toxic in brackish water and seawater than in freshwater.

Disease per se is not discussed here, although emphasis on environmental control as a means of disease prevention is the subject of later sections of the chapter. Good general information on diseases of fishes and aquatic invertebrates is available from many sources (e.g., Amlacher et al. 1970, Bullock 1971, Bullock et al. 1971, Dulin 1976, Hoffman 1967, Hoffman and Meyer 1974, Johnson 1978, Kabata 1970, Kingsford 1975, Klontz 1973, National Academy of Science 1973, Pauley and Tripp 1975, Reichenbach-Klinke and Elkan 1965, Sarig 1971, Sindermann 1977, Snieszko 1970, van Duijn 1973, Wellborn and Rogers 1966). Bibliographies were published by Conroy (1968), Johnson (1968), and Sindermann (1970). Immunity in fishes was discussed by Anderson (1974) and Lom (1969). Ribelin and Migaki (1975) edited a work on the pathology of fishes.

8.1 AMMONIA TOXICITY

Ammonia is the most toxic form of inorganic nitrogen produced in aquarium water. Ammonia in aquariums originates from the mineralization of organic matter by heterotrophic bacteria and as a by-product of nitrogen metabolism by most aquatic animals (Fig. 1-2, p. 2).

Hydrolysis

The hydrolysis of ammonia in natural waters, as shown by the reaction below, has a pK value of about 9.0, so that the percentage of ammonium

112

ion is always greater than the percentage of free ammonia. The factors affecting ammonia hydrolysis are important because many fish culturists consider NH_3 to be significantly more toxic than NH_4^+.

$$NH_4^+ + H_2O \rightleftharpoons NH_3 + H_3O^+ \tag{23}$$

Reaction 23 is controlled by pH, temperature, and salinity, with pH exerting the greatest effect. A pH increase of one unit causes the percentage of free ammonia to increase about tenfold. Rising temperature and decreasing salinity cause much smaller increases. The temperature effect is the result of increased hydrolysis of ammonium ions at higher temperature levels; the salinity effect is the result of the decreasing activity of free ammonia in solutions of increasing ionic strength (Hampson 1976). Thus at identical temperature and pH values, and the same amounts of total NH_4-N in solution, seawater contains slightly less NH_3 than freshwater. This is shown in Table 9-1 (p. 130).

Excretion and Toxicity

Most aquatic animals are *ammonotelic*, excreting more than 50% of their waste nitrogen as ammonia, primarily through the gills. For years it was thought that NH_3 could pass across tissues and that NH_4^+ could not (e.g., Milne et al. 1958, Warren 1962). This theory must now be discarded. It is clear that both forms of ammonia can penetrate tissues. Recent research indicates that NH_4^+ is the major form of ammonia excreted by the gills of both fishes and invertebrates living in waters of all salinities (Evans 1973, 1975; Kerstetter and Keeler 1976; Maetz et al. 1976; Mangum et al. 1978; Payan and Maetz 1973; Payan and Matty 1975). These investigators consider the elimination of the ammonium ion to be an exchange process in which NH_4^+ produced metabolically is exchanged at the surface of the gill for Na^+ in the environment. The importance of NH_3 excretion is less certain because little information is available.

The means by which ammonia becomes toxic must also be reconsidered. In the past, it was simple enough to think of the problem in terms of partial pressure gradients of NH_3 across the gill. As the partial pressure of NH_3 in the water increased, free ammonia simply diffused into the animal down a concentration gradient. This could happen when the level of NH_3 in the water exceeded the NH_3 concentration in the blood. Such a mechanism now seems questionable. Fromm and Gillette (1968) showed that when rainbow trout (*Salmo gairdneri*) were subjected to elevated levels of environmental ammonia, blood ammonia also increased, but blood ammonia concentrations were always greater than the concentrations of environmental ammonia on the other side of the gill. According to these authors, there is no experimental evidence demonstrating that ammonia enters fish gills from the environment. This should be taken to mean ammonia of either species. They proposed instead that elevated ammonia levels in the water caused the forced retention of metabolic ammonia, which then became toxic. Indirect proof of this hypothesis was obtained by Fromm and Gillette (1968) and Olson and Fromm (1971), who showed that the excretion rate of ammonia by rainbow trout decreased as the ammonia concentration in the water increased. Future experimentation may reveal a mechanism by which ammonia can enter the gills of aquatic animals and poison them directly. Until then, it must be assumed that high ammonia levels in the water exert a toxic effect indirectly, and that actual toxicity occurs when animals are unable to rid themselves of excessive ammonia produced during nitrogen metabolism.

The pH of aquarium water influences the toxicity of ammonia by controlling the hydrolysis of NH_4^+. The percentage of free ammonia increases in proportion to ammonium ion with increasing pH, and NH_3 is considered by many researchers and fish culturists to be the more toxic form (Armstrong et al. 1978, Burkhalter and Kaya 1977, Fromm and Gillette 1968, Hampson 1976, Robinette 1976, Rubin and Elmaraghy 1977, Schreckenbach et al. 1975). Both forms of ammonia can cross tissue barriers, as mentioned previously, and both are toxic, although NH_3 appears to be more toxic than NH_4^+ at higher pH values (Armstrong et al. 1978).

Ammonia toxicity is exacerbated by low dissolved oxygen levels, but the mechanism is obscure. Reichenbach-Klinke (1967) noted that the oxygen content in the tissues of several species of European freshwater teleosts decreased as the tissue concentration of NH_3 increased, and that resistance to ammonia was lower during anoxic conditions. Other investigators have also reported that the toxicity of NH_3 is magnified in waters of low oxygen tension (Downing and Merkens 1955; Larmoyeux and Piper 1973; Lloyd 1961; Merkens and Downing 1957; Smith 1972; Smith and Piper 1975; Wuhrmann and Woker 1948, 1953).

According to one theory, elevated levels of ammonia in the environment in some way interfere with the ability of hemoglobin to retain oxygen. Reichenbach-Klinke (1967) saw a reduction in red blood cells in freshwater teleosts after the fishes had been kept in water with sublethal concentrations of ammonia. Brockway (1950) reported that when ammonia in the water increased to about 1.0 mg NH_4–N l^{-1}, the oxygen content of rainbow trout blood decreased to about 14% of its normal value and the CO_2 content increased by approximately 15%. Fromm and Gillette (1968), on the other hand, observed that ammonia levels up to 10 mg NH_3–N l^{-1} had no significant effect on the ability of hemoglobin to combine with oxygen when the erythrocytes were suspended in Ringer and studied *in vitro*.

Sousa and Meade (1977) used spectrophotometric methods to study hemoglobin solutions from coho salmon that had been subjected to high levels of environmental ammonia. They noted a progressive shift in the absorption configuration of oxygenated hemoglobin toward deoxygenation. The authors postulated that prolonged exposure of the fish to elevated ammonia levels in the water resulted in acidemia. This, in turn, interfered with the ability of hemoglobin to transport oxygen. The acidemia brought on by upsetting the internal acid-base balance caused oxygen to be released prematurely (Bohr effect).

8.2 NITRITE TOXICITY

Nitrite in the blood oxidizes hemoglobin to methemoglobin, which is incapable of transporting oxygen (Jaffe 1964). Methemoglobin in fishes can be detected by the color of the blood and gills, which turn brown. It has been thought in the past that the presence of methemoglobin could be correlated with the mortality rate in aquatic animals, but this may not always be true, particularly when percent mortality in freshwater and seawater are compared using the same species of bioassay animal. Crawford and Allen (1977) showed that in chinook salmon fingerlings subjected to levels of 8.18 mg NO_2-N l^{-1} in freshwater, 44% methemoglobin occurred with 70% mortality. Chinook salmon in seawater exposed to 246.97 mg NO_2-N l^{-1} had 74% methemoglobin with only 10% mortality.

Table 8-1 summarizes some of the published values for nitrite toxicity in freshwater, brackish water, and seawater. As shown, mortality decreases markedly with increasing salinity of the medium. Crawford and Allen (1977) attributed this to the possible protective effect of increased calcium ions, but conceded that perhaps another ion present in the medium also was involved. Perrone and Meade (1977) offered good evidence that chloride exerts a protective effect. No mortality occurred in their test animals (yearling coho salmon) in 48 hr at nitrite concentrations of 29.8 mg NO_2-N l^{-1} and chloride levels of 261.3 mg Cl l^{-1}. Other salmon exposed to 3.8 mg NO_2-N l^{-1} and 2.5 mg Cl l^{-1} showed 58.3% mortality in 12 hr. No protective mechanism has been postulated, but perhaps cations combine with the nitrite ion and prevent its uptake from solution. Chloride ion could compete directly with nitrite ion, and its greater concentration in solution might inhibit the assimilation of NO_2^-. At any rate, it is doubtful whether nitrite poses a serious threat to aquarium animals maintained in brackish water or seawater aquariums, although its presence in low concentrations is definitely hazardous in freshwater.

8.3 NITRATE TOXICITY

Nitrate is not acutely toxic to aquatic animals even in large concentrations, although its effects over extended periods of time have not been

Table 8-1. Published Lethal Concentrations of NO_2-N in Freshwater (FW), Brackish Water (BW), and Seawater (SW)

Species and Size or Age	Concentration (mg NO_2-N l^{-1})	Time (hr)	% Mortality	Medium	Source*
Rainbow trout (yearling)	0.55	24	55	FW	a
Chinook salmon (32 g)	0.50	24	40	FW	a
Chinook salmon (fingerling)	0.88	96	50	FW	b
Rainbow trout (12 g)	0.19	96	50	FW	c
Rainbow trout (9.1 g)	0.23	96	50	FW	d
Coho salmon (yearling)	3.80	12	58.3	FW	e
Chinook salmon (fingerling)	19.00	48	50	FW	f
Malaysian prawn (larvae)	8.60	96	50	BW	g
Chinook salmon (fingerling)	1070.00	48	10	SW	f
American oyster (adult)	658.00	96	50	SW	h
American oyster (juvenile)	798.00	96	50	SW	h
Quahog (adult)	1190.00	96	50	SW	h
Quahog (juvenile)	1133.00	96	50	SW	h

* a, Smith and Williams (1974); b, Westin (1974); c, Russo et al. (1974); d, Brown and McLeay (1975); e, Perrone and Meade (1977); f, Crawford and Allen (1977); g, Armstrong et al. (1976); h, Epifanio and Srna (1975).

determined. It is possible that the pale gills seen in captive fishes after several months are the result of elevated nitrate, but this is only speculation.

Knepp and Arkin (1973) reported that nitrate levels of 400 mg NO_3-N l^{-1} did not affect the percent mortality or growth of largemouth bass (*Micropterus salmoides*) and channel catfish (*Ictalurus punctatus*), both of which are freshwater species. Higher concentrations were not tested. Westin (1974) determined that nitrate was 2000 times less toxic than nitrite to chinook salmon (*Oncorhynchus tshawytscha*) and rainbow trout maintained in freshwater. Signs of trauma in chinook salmon did not occur until 5–8 days when the nitrate concentration was 1000 mg NO_3-N l^{-1} or slightly less. These signs included the inability to swim upright, labored ventilation, and reduced movement interspersed with erratic swimming. Some of the fishes had pale pink to dark red-brown gill filaments. Another interesting observation was that nitrate proved to be slightly more toxic when the animals were kept in brackish water ($S = 15‰$). In salmon, toxicity of nitrate in brackish water was 1.14–1.41 times greater than in freshwater. In trout acclimated to brackish water, nitrate toxicity proved to be 1.24–1.38 times greater than in freshwater.

Hirayama (1966b) observed increased ventilation rates in the common octopus (*Octopus vulgaris*) maintained in seawater with low pH and alkalinity values and high nitrate concentrations. Death occured at 7 hr when the nitrate level was 1400 mg NO_3-N l^{-1}. Kuwatani et al. (1969) concluded that nitrate concentrations up to 1000 mg NO_3-N l^{-1} had no effect on growth of the Japanese pearl oyster (*Pinctata fucada*), so long as the pH of the seawater was higher than 8.05.

Epifanio and Srna (1975) demonstrated that large amounts of sodium nitrate were not toxic to two species of bivalve mollusks kept in low salinity seawater ($S = 27‰$). The 96-hr LC_{50} levels of sodium nitrate were 2604 mg NO_3-N l^{-1} for adult American oysters (*Crassostrea virginica*) and 3794 mg NO_3-N l^{-1} for juveniles. In the quahog (*Mercenaria mercenaria*), 50% mortality did not occur even when the $NaNO_3$ concentration reached nearly 75% of the total salinity, and the authors considered it unnecessary to test higher concentrations.

8.4 HEAVY METAL TOXICITY

The common heavy metals (mainly Cr, Pb, Hg, Cu, and Zn) are present in water in trace amounts. All are toxic to aquatic animals in greater, but still very low, concentrations. Many are lethal in quantities less than 1.0 mg l^{-1}. Zinc and copper are the two heavy metals that will be considered here, because they are sometimes used to treat protozoan infections in fishes, particularly marine fishes.

The toxic effects of zinc or copper compounds added to aquarium water are reduced after a time by precipitation of the heavy metal ion with carbonate. Thus the effectiveness by which heavy metals are precipitated in water is a direct function of alkalinity, and not the hardness or cation concentration. Both zinc and copper in dissolved form cause fishes to produce excessive mucus. The mucus is shed into the water along with the encysted stages of some parasites. It is thought that the swarming stages of such parasitic protozoans as *Cryptocaryon irritans* are more susceptible to heavy metal poisoning and that treating an aquarium with zinc or copper kills these organisms before they can attach to the host. However, this has never been demonstrated to happen *in vitro*.

Increased ventilation rate is common in fishes exposed to toxic concentrations of copper and zinc (Jones 1938, 1939; Sellers et al. 1975). Several investigators suggested that heavy metals cause tissue hypoxia by interfering with normal gas exchange (Ellis 1937, Skidmore 1964, Vernon 1954). Burton et al. (1972) concluded that the ultimate cause of death in copper-treated fishes was tissue hypoxia brought about by the coagulation or precipitation of mucus on the gills, and by the accompanying cytological damage to gill tissue. Histopathological changes in the gill tissue of fishes exposed to heavy metals were observed in some reports. Baker (1969) noted that the effects of low copper levels to the winter flounder (*Pseudopleuronectes americanus*) were an extracted appearance of the gill lamellae, vacuolated epithelial layers, and a reduction in the number of lamellar mucus cells. Matthiessen and Brafield (1973) mentioned the coalescing of secondary lamellar epithelia in threespine sticklebacks (*Gasterosteus aculeatus*) exposed to zinc.

Heavy metals produce other harmful effects, including disruption of enzyme activity (Christensen 1971/1972, Jackim et al. 1970), impaired ionic regulation (Lewis and Lewis 1971), hyperglycemia or elevated blood glucose (Watson and McKeown 1976), and alteration of the normal blood characteristics (Cardeilhac and Hall 1977, McKim et al. 1970). Gardner and LaRoche (1973) reported that exposure to copper produced cellular changes in the mechanoreceptors of the lateral line canals in the heads of adult mummichogs (*Fundulus heteroclitus*) and Atlantic silversides (*Menidia menidia*), in addition to lesions in the olfactory organs and kidney, and hemmorhage in the brain and periorbital connective tissue.

8.5 IMMUNITY AND THE ENVIRONMENT

There is a voluminous literature on disease treatments of aquatic animals, whereas comparatively little has been written on disease prevention by means of environmental control. This section is an attempt to reconcile the cause-versus-effect aspects of infectious disease as they relate to environmental conditions. Admittedly, any such attempt is bound to be incomplete because the relationship between disease and environmental stress is not clearly defined.

Evidence has been presented in previous chapters that aquatic animals are acutely sensitive to slight changes in water quality. It should be emphasized again that even though most animals can control their internal responses to these changes to some extent, any response is likely to be magnified under captive conditions. In the wild, the natural dispersal of both hosts and parasites allows a host species to survive even though numerous individuals may perish when other unfavorable conditions prevail. But in captivity, dense crowding facilitates parasite transfer. When the water is not properly managed, parasites may proliferate and eventually overwhelm their hosts. Under such conditions, any natural resistance the host may have is quickly broken down by the sheer number of infectious agents.

Infectious agents have been found in all aquatic animals, but in aquatic animal culture only transmissible forms are significant. The severity of an infection depends largely on the physiological well-being of the host. This, in turn, depends directly on environmental factors. In healthy animals infections are often *latent*—the parasites are present but not in infectious, or disease-producing, stages. Most epizootic outbreaks of disease encountered in aquariums are caused by bacteria and protozoans. These organisms normally can be held in check by stringent quality control of the environment.

The virulence of helminth and arthropod parasites is less dependent on the water quality factors to which the host is subjected. Heavy infestations often occur on otherwise healthy animals despite carefully controlled factors such as temperature, dissolved oxygen, and ammonia. The infested animals usually require treatment.

Parasitic bacteria, protozoans and, to a lesser extent, viruses are more troublesome in aquariums than are helminths and arthropods. The latter organisms can be kept out of aquarium water by the methods described in the next section. However, bacteria and protozoans usually are present on the gills and external surfaces of fishes and other aquatic animals. The physiological condition of the host determines whether they remain latent or become infectious. A temporary decline in dissolved oxygen weakens the host; so does an increase in free CO_2, ammonia, and perhaps DOC. Heterotrophic bacteria proliferate as the total organic carbon concentration increases and normally harmless forms may become infectious when present in large numbers. Temperature fluctuations cause considerable stress to the host and often result in chronic protozoan infections even long after the temperature has returned to normal.

Bacterial and protozoan infections often occur when new animals are added to a conditioned aquarium. It is commonly thought that the infectious agents enter with the new animals and are transmitted to the established ones, but the problem may not be this simple. For instance, if the established animals are carrying the same infectious organisms as the newcomers, but only in the latent stages, the addition of the new animals may cause these latent forms to become infectious. This can be brought about by several factors, including a rise in ammonia during the temporary shift in the carrying capacity of the filter bed. There may also be a rise in DOC and, in poorly buffered water, an increase in free CO_2. In other words, water quality may still be the underlying cause even though the only visible change has been the addition of more animals. As shown in Fig. 8.1 a decline in water quality resulting in lowered disease resistance of the host is enough to alter the virulence of a latent parasite.

The mucus on the outer surfaces of aquatic animals serves as the first line of defense against the invasion of ectoparasites. It forms a protective

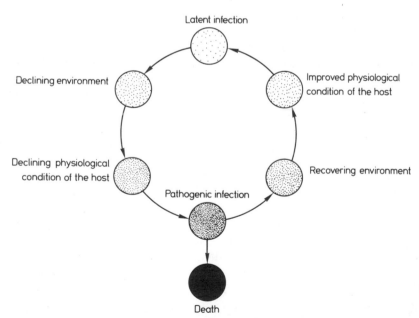

Figure 8-1. The virulence of infectious agents in relation to environmental conditions. Original.

sheath that helps to maintain the host-parasite balance in favor of the host. There is evidence that fish mucus contains antibodies that repel parasitic protozoans (Lom 1969). The synthesis of antibodies by cold-blooded animals is temperature dependent. Antibody production in fish mucus is inhibited above or below specific temperatures, depending on the species. Lom (1969) cited evidence that raising the temperature of the culture water to 20°C caused trichodinids to drop off carp because " . . . the protective capacity of the mucus is manifested only at the elevated temperature . . . ." At 10°C the fish were unable to rid themselves of the parasites, even though the growth of the trichodinids was inhibited at the lower temperature.

Temperature fluctuations and accumulating toxic metabolites are two of the primary factors causing epizootic outbreaks of infectious bacteria and protozoans.

8.6 DISEASE PREVENTION

Four factors are involved in environmental disease prevention: (1) maintaining proper environmental conditions (including water quality and stable temperatures); (2) sterilizing the circulating aquarium water; (3) providing adequate nutrition; and (4) preventing the introduction of infectious agents from outside sources. The first two factors have been discussed in previous chapters, and the third is not within the scope of this book. The fourth factor is considered here.

Raw Water

Many infectious agents are present in unprocessed natural, or raw, water. Raw water should never be pumped directly from a natural source into a culture system. Makeup for partial water changes should be prefiltered with rapid sand filters to remove the initial turbidity. It should then be filtered with diatomaceous earth. DE filters are effective in removing any bacteria and protozoans that may have passed through the sand beds. After filtration with DE, the water should be kept in darkened storage vats for 2 weeks and moderately aerated. Most parasites die within this time when no hosts are available. This means that at least two vats should be available, each holding enough water for the standard 10% change routinely provided for each aquarium biweekly. There should also be enough extra volume in each vat to start up a new aquarium in case of emergency.

In closed-system culturing, prefiltration of raw water is carried out intermittently and the filters are not in continuous operation. The prefiltration system (rapid sand and DE filters in series) should be designed so that the filters can be recycled independently of the storage vats and aquariums. This arrangement is shown diagrammatically by dotted lines in Fig. 8.2. The position of the sterilizer (ozone or UV) is also shown in the figure. The water should be sterilized as it leaves the vats on its way to the aquariums after the 2-week aging period.

The prefiltration system should be chemically sterilized immediately after processing a batch of raw water, then drained and left dry until the next batch is filtered. Sterilization with UV irradiation or ozone is not effective in this case, because neither kills infectious organisms deep in the sand beds or attached to the DE filter elements, walls, and bays. Superchlorination is the most effective technique, although care should be taken to remove all traces of residual chlorine.

Superchlorination of Prefilters

1. Adjust the valves so that the prefilters operate as a closed system (recycle).
2. Backwash the DE filter. Do not precoat.
3. Add sodium hypochlorite until the level of free chlorine is at 50 mg free Cl_2l^{-1} (see tests in American Water Works Association et al. 1976).
4. Turn on the pumps and recycle the chlorinated water through the filters for 2 hr.

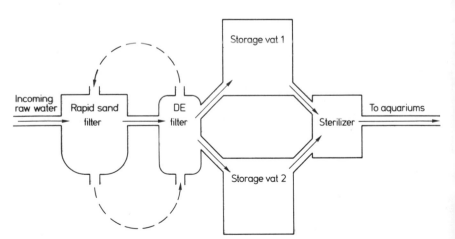

Figure 8-2. A prefiltering arrangement for processing large volumes of influent water. Original.

5. Reduce the free chlorine to zero with sodium thiosulfate ($Na_2S_2O_3$). The amount required will vary and must be determined by trial and error.

6. Drain the filters and leave them dry until the next batch of raw water is prefiltered.

7. When starting prefiltration again, let the system flush to waste for 30 min before diverting the water to the storage vats.

Infected Animals

New stocks of animals should not be added to the main water systems until they have been isolated (quarantined) and have proved to be free of infections (microorganisms) or infestations (macroorganisms) for 4 weeks.

Live Food

Live food may be unsafe unless it is tank raised. Cultured brine shrimp are suitable to feed to both freshwater and seawater animals. Forms of live food collected in the wild are potential carriers of parasites. Live food collected from marine waters should not be fed to brackish water or seawater animals; freshwater food organisms should not be fed to fresh water animals unless they are tank raised. However, parasites normally present on marine food organisms can seldom survive immersion in fresh water and the latter are safe to feed to freshwater animals. The reverse is also true, and freshwater forms, such as wild *Daphnia,* are safe to feed to seawater animals. Freshwater minnows should never be fed to carnivorous freshwater fishes, although they are safe to feed to seawater species. In all instances, commercially prepared foods are preferable to live or frozen animal flesh from the standpoint of disease transfer.

8.7 TREATING DISEASES

Without the proper controls, treating diseases and at the same time maintaining stable environmental conditions are antagonistic processes. The aquarist who is too quick with his beaker of remedies may aggravate an already serious situation by killing beneficial microbes in the filter bed. When this happens, the decline in water quality accelerates and still more factors combine to plague the animals. With this in mind, any treatment program must be approached thoughtfully and with the realization that failure is imminent in a deteriorating environment. In unskilled hands, the "cure" may be more lethal than the disease.

In most cases, it is best not to treat animals infected by bacteria or protozoans. Loss of stocks to these organisms, as previously mentioned, can often be traced to an environmental factor. The most heavily infected stocks must be removed and the factor responsible corrected. Animals with mild infections and in otherwise good health then stand a good chance of recovery.

Helminth and arthropod parasites have greater resistance to the immune responses of the host and cannot be handled in the same manner. The only sure way to eradicate these organisms is to remove all the animals and superchlorinate the aquarium tank, and filtration system. A free chlorine level of 50 mg free Cl_2 l^{-1} sustained for 2 hr is sufficient. The chlorine level must then be reduced to zero with sodium thiosulfate and the system drained and flushed thoroughly with tap water. The aquarium can then be refilled with newly processed water and conditioned again with new stocks that have been quarantined and acclimated. This technique is admittedly drastic, but it is also effective. Heavy infestations of helminths and arthropods are seldom a problem if quality control procedures, such as the ones outlined earlier, are followed stringently.

The actual selection of medication has been the subject of several books, so nothing specific is mentioned here. In selecting a compound, however, three things should be kept in mind. First, one should be chosen that is as specific as possible for the infectious agent in question. Second, compounds that cause aquatic animals to shed copious quantities of mucus should be avoided. When the mucus is destroyed the animals are left unprotected against secondary invasion by other pathogens, or even against reinfection. Heavy metals, therefore, should be used sparingly. Third, diseased animals should never be treated directly in the aquarium. Some chemotherapeutic compounds interrupt nitrification (Section 1.2). For this reason, treatment tanks equipped with biological filtration are seldom fully functional.

8.8 MANAGEMENT PRACTICES

Evidence supporting NH_3 as being significantly more toxic than NH_4^+ has been obtained by means of bioassay techniques using whole animals (e.g., Lloyd and Orr 1969, Wuhrmann and Woker 1948). Definitive proof must wait until the same differences in toxicity are shown to take place at a cellular level. No matter what the results, they will be of academic interest only, because the factors that control ammonia hydrolysis in aquarium water (pH, temperature, and salinity) cannot be manipulated easily without causing other harmful effects. For example, even if ammonia is less toxic at low pH values, this is of little solace to marine aquarists. Wet-thumb experience has demonstrated that seawater animals live better at pH values of 8.0 or higher, although the reasons why are not known. Perhaps this is the ideal pH range to sustain efficient transfer processes across the gills and integuments. Temperature and salinity also must be tightly controlled for reasons that are just as important. If water temperature is altered drastically, the enzyme systems of ectothermic animals are affected, and the ionic strength of a freshwater aquarium cannot be raised simply because the percentage of free ammonia will be reduced.

Temperature, pH, and salinity must be controlled according to the known ecological and physiological requirements of the animals and plants being maintained. Heavy aeration appears to reduce the toxic effects of ammonia to aquatic animals. The removal of uneaten food, dead animals and plants, routine administering of partial water changes, and keeping animal and plant densities at moderate levels are the most effective means of preventing high ammonia levels in solution. The upper limit should not exceed 0.1 mg total NH_4-N l^{-1}.

Nitrite appears to be acutely toxic only in waters of low ionic strength, but this observation is based on limited data. As such, the acceptable upper limit in all aquarium waters is 0.1 mg NO_2-N l^{-1}. The long-term effects of nitrate are unknown, although its short-term effects can be discounted (i.e., nitrate is not acutely toxic). It is difficult to believe that high nitrate levels do not have some harmful effect and efforts should be made to keep the concentration below 20.0 mg NO_3-N l^{-1}. This can be accomplished by a program of regular partial water changes, and by culturing plants in the water system and harvesting some of them periodically.

Evidence points to the treating of fishes with heavy metals as a dangerous practice of doubtful therapeutic value. In most cases, the unfavorable effects on the host outweigh any benefits gained by killing its parasites. No one has yet established the lethal limits of zinc or copper for any species of marine teleost. Treating a mixed-species aquarium is particularly ludicrous because the long-term effects, and even the lethal thresholds among fishes in general, are known to differ considerably. Moreover, no one has yet shown the effects of a heavy metal on every stage in the life cycle of a single parasitic organism. In short, the treatment may do more harm to the host than to the parasite.

The percentage of zinc or copper that remains soluble after being exposed to aquarium seawater is difficult to predict. A heavy accumulation of detritus on the surface of a filter bed, in addition to a high concentration of dissolved organic carbon in the water, affects the copper or zinc levels that can be tolerated by microorganisms and aquarium animals and plants.

There is evidence that organic matter in activated sludge combines with copper and reduces its toxicity to nitrifying bacteria (Tomlinson et al. 1966). Chelation apparently renders copper less toxic to freshwater fishes (Doudoroff and Katz 1953). Fitzgerald (1963) showed that chelating cupric sulfate with citric acid in lake water rendered it 500 times less toxic to bluntnose minnows (*Pimephales notatus*). Chelation enabled 90% of the copper to remain soluble at pH values of both 6.0 and 8.5.

Chelated copper and zinc compounds often are used to treat seawater aquariums when prolonged soluble levels are desired. However, heterotrophic bacteria may be resistant to the toxic effects of heavy metal chelates and decompose the organic substrate. This allows the heavy metal ions to combine with carbonate and precipitate. It is significant that none of the published dosage levels recommended for treating seawater aquariums mention DOC as a factor that might alter the toxic state of the metal ion. Chelation, as just noted, reduces the toxicity of heavy metals to fishes. Presumably, its effects on the parasites are similar (Spotte 1970).

The primary mechanism of heavy metal toxicity is impairment of respiration and tissue hypoxia. Any treatment of the water with copper or zinc should be accompanied by heavy aeration. Diseased fishes should be removed from the aquarium and treated in separate containers that do not contain calcareous gravel or water that is high in DOC. Heavy metals that precipitate in the aquarium tank itself may not be removable by conventional methods (e.g., partial water changes, physical adsorption). Water treated with heavy metals should not be subjected to physical adsorption by activated carbon or foam fractionation, because both techniques quickly remove chelated metals from solution. Ozonation may also decompose organic compounds containing heavy metals, freeing the metal ions to combine with carbonate and precipitate.

CHAPTER 9

Analytical Methods

Besides the usual glassware, a laboratory for the routine analysis of aquarium water should be equipped with a spectrophotometer (not a colorimeter), pH meter (preferably with an expanded scale), desiccator, magnetic stirrer, water still, deionization column, drying oven, analytical balance, bunsen burner, and refrigerator with a freezer compartment.

The tests given here are adequate to maintain aquarium water in good condition. Detailed discussions of analytical methods are given in other books. An excellent handbook for freshwater analysis was written by Stainton et al. (1977). For complete discussions of the practical and theoretical aspects of seawater analysis see Grasshoff (1976), Riley et al. (1975), Smith and Windom (1972), and Strickland and Parsons (1972). The use of specific ion electrodes for phosphate and nitrate measurements in freshwater were discussed by Tanaka et al. (1977) and Shechter and Gruener (1976), respectively. Barica (1973) published a report on the use of specific ion electrodes for ammonia determination in freshwater aquariums; use of these instruments for ammonia analysis in seawater aquariums was covered by Gilbert and Clay (1973) and Srna et al. (1973).

9.1 REACTIVE PHOSPHORUS (AS TOTAL PO$_4$-P)

A procedure for determination of reactive phosphorus (also called inorganic phosphorus, orthophosphate, and phosphate) is given here, despite the fact that phosphorus was only briefly mentioned in the text. Values in aquarium water ordinarily range from about 0.5 to 6.0 mg PO$_4$-P l^{-1}. The method described is adapted from Murphy and Riley (1962) and Martin (1972) and works well in waters of all salinities.

Preparation of Glassware

1. Start with new glassware.
2. Add 8.0 ml of mixed reagent (see Reagents) for each 100 ml of volume and swirl to coat the walls of the glassware.

127

3. Allow to stand for 20 min.

4. Discard the mixed reagent and rinse each container thoroughly with triple-distilled water.

5. Store glassware in a 0.1% sulfuric acid solution (1.0 g concentrated H_2SO_4 l^{-1} of distilled water).

6. Glassware can be used again without repeating steps 1 through 4, provided it has not been washed in detergent between tests.

Sampling

Water samples should be collected in glass 250-ml flasks prepared by the procedure above and covered immediately with new aluminum foil. Analyze within 2 hr, if possible. Murphy and Riley (1962) reported that samples could be preserved by addition of chloroform (1.0 ml 150 ml^{-1}), although others have found the technique ineffective (see Martin 1972). Samples can be stored in polyethylene containers if they are quick-frozen. Hassen-Teufel et al. (1963) and Heron (1962) observed loss of phosphate from polyethylene containers at room temperature.

Reagents

Sulfuric acid solution (5.0 N).   Add 70.0 ml of concentrated H_2SO_4 (96%, specific gravity 1.84) to about 500 ml of triple-distilled water and dilute to 1.0 l. *Use caution.*

Ammonium molybdate solution.   Dissolve 40.0 g of reagent grade ammonium molybdate ($[NH_4]_6Mo_7O_{24} \cdot 4H_2O$) in about 900 ml of triple-distilled water and dilute to 1.0 l. Store in a polypropylene bottle in the dark, or in an amber, phosphate-free borosilicate glass bottle. (1 hr)

Potassium antimonyl tartrate solution.   Dissolve 1.4 g of reagent grade material (tartar emetic, $K[SbO]C_4H_4O_6 \cdot 0.5H_2O$) in triple-distilled water and dilute to 500 ml. Store in a phosphate-free glass bottle. (many months)

L-ascorbic acid solution.   Dissolve 1.3 g of reagent grade material in 75.0 ml of triple-distilled water.

Mixed reagent solution.   Add 125 ml of sulfuric acid (5.0 N) to a 250-ml erlenmeyer flask and add 12.5 ml potassium antimonyl tartrate solution. Mix thoroughly. Add 37.5 ml ammonium molybdate solution and mix thoroughly. Add 75.0 ml L-ascorbic acid solution and mix again. (24 hr)

Standard phosphate solution (100mg PO_4-P l^{-1}).   Dissolve 0.556 g of reagent grade anhydrous (oven dried at 105°C for 3 hr) potassium dihydrogen phosphate (KH_2PO_4) in distilled water and dilute to 100 ml. Mix thoroughly. Dilute 10 ml of this solution to 1.0 l with distilled water and add

1.0 ml of chloroform to retard the growth of bacteria. Store in an amber, phosphate-free borosilicate glass bottle. (many months)

Procedure

1. Warm the sample to room temperature, then filter through a 0.45-μm membrane filter.

2. Pipette a volume of sample likely to contain 0.1 mg PO$_4$-P into a 50-ml volumetric flask, add 8.0 ml of mixed reagent, dilute to 50 ml with distilled water, and mix.

3. Prepare a blank by adding a similar volume of sample to a 50-ml volumetric flask, dilute to 50 ml with distilled water, and mix.

4. Pipette 1.0 ml of standard (0.1 mg PO$_4$-P) into a 50-ml volumetric flask, add 8.0 ml of mixed reagent, dilute to 50 ml with distilled water, and mix.

5. Using the blank and a 2- or 5-cm cell, set the absorbance to 0.00 at 690 nm.

6. After 10 min, but before 3 hr, read and record the absorbance of the standard at the above settings.

7. Read and record the absorbance of the sample.

8. If the absorbance is greater than 1.00, prepare a new solution (step 2) using a smaller volume. If the absorbance is below 0.2, use a larger volume. Also prepare a new blank, but not a new standard.

9. The concentration of PO$_4$-P in mg l^{-1} is given by

$$c_u = \frac{c_s V_s A_u}{V_u A_s} \tag{24}$$

where c_u = the concentration of the unknown in mg PO$_4$-P l^{-1}, c_s = the concentration of the standard in mg PO$_4$-P l^{-1}, A_u = absorbance of the unknown, A_s = absorbance of the standard, V_u = volume of the unknown pipetted in litres, and V_s = volume of the standard pipetted in litres.

9.2   AMMONIA (AS TOTAL NH$_4$-N)

Ammonia is determined as total NH$_4$-N, or "ammonia nitrogen." To obtain the concentration of ammonium ion (NH$_4^+$), multiply the measured value by 1.3. The percentage of the total NH$_4$-N that exists as NH$_3$ can be read from Table 9-1 (freshwater and full-strength seawater), provided that temperature, pH, and salinity measurements are made when the water sample is collected. More complete tables for freshwater were published by Emerson et al. (1975) and Trussell (1972), and for saline waters by

Table 9-1.   Percentage of Free Ammonia (as NH_3) in
Freshwater (FW) and Seawater (SW) at Varying pH and
Temperature. Salinity = 34‰ at an Ionic Strength of 0.701 m
for the Seawater Values. Freshwater Data from Trussel (1972);
Seawater Values from Bower and Bidwell (1978)

pH	10°C		15°C		20°C		25°C	
	FW	SW	FW	SW	FW	SW	FW	SW
7.0	0.19		0.27		0.40		0.55	
7.1	0.23		0.34		0.50		0.70	
7.2	0.29		0.43		0.63		0.88	
7.3	0.37		0.54		0.79		1.10	
7.4	0.47		0.68		0.99		1.38	
7.5	0.59	0.459	0.85	0.665	1.24	0.963	1.73	1.39
7.6	0.74	0.577	1.07	0.836	1.56	1.21	2.17	1.75
7.7	0.92	0.726	1.35	1.05	1.96	1.52	2.72	2.19
7.8	1.16	0.912	1.69	1.32	2.45	1.90	3.39	2.74
7.9	1.46	1.15	2.12	1.66	3.06	2.39	4.24	3.43
8.0	1.83	1.44	2.65	2.07	3.83	2.98	5.28	4.28
8.1	2.29	1.80	3.32	2.60	4.77	3.73	6.55	5.32
8.2	2.86	2.26	4.14	3.25	5.94	4.65	8.11	6.61
8.3	3.58	2.83	5.16	4.06	7.36	5.78	10.00	8.18
8.4	4.46	3.54	6.41	5.05	9.09	7.17	12.27	10.10
8.5	5.55	4.41	7.98	6.28	11.18	8.87	14.97	12.40

Bower and Bidwell (1978). The procedure given below is adapted from
Solórzano (1969) and gives good results in waters of all salinities. Procedures that use Nesslerization do not work in saline waters without modification (Manabe 1969).

Sampling

Water samples should be collected in clean glass 250-ml flasks and
covered immediately with new aluminum foil. The samples should be
analyzed within 2 hr. Degobbis (1973) reported that seawater samples could
be preserved for up to 2 weeks by the addition of phenol at concentrations
of 0.4 g 100 ml^{-1}. The use of chloroform as a preservative proved unsuitable, as did slow and quick freezing. Storage in glass containers without
preservatives resulted in a significant increase in ammonia; storage in
polyethylene containers caused the ammonia level to decrease significantly.

Reagents

Phenol-alcohol solution. Dissolve 20.0 g of reagent grade phenol in 200 ml of 95% vol vol^{-1} ethyl alcohol (USP).

Sodium nitroferricyanide (0.5%). Dissolve 1.0 g of sodium nitroferricyanide in 200 ml of deionized water. Store in an amber bottle and refrigerate. (1 month)

Alkaline solution. Dissolve 100 g of trisodium citrate and 5.0 g of sodium hydroxide in 500 ml of deionized water.

Sodium hypochlorite solution. Use a commercial hypochlorite solution (e.g., Clorox®) that is at least 1.5 N.

Oxidizing solution. Mix 100 ml of sodium citrate solution and 25 ml of hypochlorite solution. (8 hr)

Standard ammonia solutions. Dissolve 0.472 g of $(NH_4)_2SO_4$ l^{-1} of deionized water to give a 100-mg NH_4-N l^{-1} solution. To 990 ml of deionized water, add 10 ml of the 100-mg solution to give a 1.0 mg NH_4-N l^{-1} solution. Use a 1.0-l volumetric flask. To 45 ml of deionized water, add 5.0 ml of the 1.0 mg solution to give a 0.1 mg NH_4-N l^{-1} solution.

Procedure

1. Add 40.0 ml of sample to a 50-ml volumetric flask.
2. Add 2.0 ml of phenol-alcohol solution from a pipette.
3. Swirl the flask and add, in order, 2.0 ml of sodium nitroferricyanide solution and 5.0 ml of oxidizing solution. Swirl the flask after each addition. Fill to the mark with deionized water and mix well.
4. Prepare a blank with 40.0 ml of deionized water according to steps 2 and 3.
5. Pipette in 40.0 ml of standard containing 0.1 mg NH_4-N l^{-1}. Follow steps 2 and 3.
6. Cover the flask to prevent contamination by atmospheric ammonia. Let stand for 1 hr at room temperature.
7. Fill a cell of thickness greater than 1.0 cm with blank and set the absorbance at 0.00 for 640 nm. The turbidity blank contains 40 ml of sample diluted to 50.0 ml with deionized water. Use this blank if necessary.
8. Read the absorbances of the standard, unknown, and turbidity blank.
9. Calculate the concentration of total NH_4-N as follows:

$$c_u = \frac{c_s V_s (A_u - A_t)}{V_u A_s} \qquad (25)$$

where c_u = the concentration of the unknown in mg NH_4-N l^{-1}, c_s = the concentration of the standard in mg NH_4-N l^{-1}, V_u = volume of unknown pipetted in millilitres, V_s = volume of standard pipetted in millilitres, A_u = absorbance of the unknown, A_s = absorbance of standard, and A_t = absorbance of the turbidity blank.

10. If the absorbance of the unknown is too high, a new sample solution and turbidity blank can be prepared using a smaller volume.

9.3   NITRITE (AS NO_2-N)

Nitrite is determined as NO_2-N, or "nitrite nitrogen." To obtain the concentration of nitrite ion (NO_2^-), multiply the measured value by 3.3. The procedure given here is adapted from Strickland and Parsons (1972) and gives reproducible results in waters of any salinity.

Sampling

Samples should be collected in clean glass 250-ml flasks and analyzed within 5 hr. Store samples in a refrigerator while awaiting analysis.

Reagents

Sulfanilamide solution.   Dissolve 5.0 g of sulfanilamide in a solution of 50.0 ml of concentrated hydrochloric acid and about 300 ml of distilled water. Dilute to 500 ml (several months).

N-(l-naphthyl)-ethylenediamine dihydrochloride solution.   Dissolve 0.5 g of dihydrochloride in 500 ml of distilled water. Store in the dark and discard when the solution turns brown.

Standard nitrite solutions.   Dry pure sodium nitrite at 110°C in a dessicator for several hours and cool. Dissolve 0.493 g of the dried material in deionized water and dilute to 1.0 l to make a 10 mg NO_2-N l^{-1} solution. Add a few drops of chloroform to retard growth of bacteria.

Low Concentration Working Standard.   Dilute the standard nitrite solution by adding 1.0 ml to a litre of deionized water. Each millilitre contains 0.01 mg NO_2-N l^{-1}. This solution should be made fresh each time.

Procedure

1. Add 2.0 ml of each of sulfanilamide solution and N (l-naphthy-ethylenediamine dihydrochloride solution to 40.0 ml of sample in a 50-ml volumetric flask. Fill to mark with deionized water.

2. Prepare standards by pipetting into 50-ml volumetric flasks 5.0, 10.0, 20.0, and 40.0 ml of low concentration working standard as in step 1. Use deionized water to prepare a reagent blank.

3. After 10 min, but before 2 hr, read absorbance against the blank at 543 nm in a cell 1 or 2 cm thick.

4. If absorbance is above 1.5, a smaller sample should be used.

5. Plot mg NO_2-N versus absorbance for the standard solutions. Read mg NO_2-N of unknown from the curve and divide this value by the volume in litres to obtain the concentration in mg NO_2-N l^{-1} (divide by 0.04 for 40.0 ml of sample).

9.4   NITRATE (AS NO_3-N)

Nitrate is measured as NO_3-N, or "nitrate nitrogen." To obtain the concentration of nitrate ion (NO_3^-), multiply the measured value by 4.4. The hydrazine reduction procedure given here works well in waters of any salinity. The original reference is Mullin and Riley (1955).

Sampling

Same as in nitrite test (Section 9.3).

Reagents

Acetone.   Reagent grade.

Buffer solution.   Mix 20.0 ml of sodium hydroxide solution with 20.0 ml of phenol solution.

Copper sulfate solution.   Dissolve 100 mg of cupric sulfate pentahydrate in 1.0 l of deionized water.

Hydrazine sulfate solution.   Dissolve 3.625 g of hydrazine sulfate in 500 ml of deionized water. (2 months)

N-(1-naphthyl)-ethylenediamine dihydrochloride solution.   Same as in nitrite test (Section 9.3).

Phenol solution.   Dissolve 9.072 g of phenol in 200 ml of deionized water. (2 months)

Reducer solution.   Mix 20.0 ml of cupric sulfate solution with 20.0 ml of hydrazine sulfate solution.

Sodium hydroxide solution.   Dissolve 28.35 g of sodium hydroxide in deionized water. Cool and dilute to 2.0 l. (6 months)

Sulfanilamide solution.   Same as in nitrite test (Section 9.3).

Procedure

1. Add 0.5 ml of sample to a 50-ml volumetric flask. Add 20.0 ml of deionized water, 2.0 ml of buffer solution, and 1.0 ml of reducer solution.
2. Let stand at room temperature for 20 hr. Shield from light.
3. Add 2.0 ml each of N-(1-naphthyl)-ethylenediamine dichloride solution and sulfanilamide solution and fill to mark with deionized water. After 10 min, but before 2 hr, read absorbance against a reagent blank at 543 nm in a cell 1 or 2 cm thick.
4. Read milligrams NO_3-N from a standard curve plotted for the nitrite test (Section 9.3) and divide by the sample volume in litres to obtain mg NO_3-N l^{-1} (divide by 0.0005 for a 0.5-ml sample).

9.5 ALKALINITY

Alkalinity can be defined as the number of equivalents of strong acid required to titrate 1.0 l of water to the CO_2-HCO_3^- end point. This is the same as the sum of the concentrations of the anions of carbonic and other weak acids. The procedure given here is a modified standard titration technique.

Reagents

0.1 N HCl solution. Use commercial standard 0.1 *N* HCl. Add NaCl until the salinity is about equal to the salinity of the sample.
pH 4.0 buffer solution. Prepare a pH 4.0 buffer solution using NBS buffer tablets or use a commercial pH 4.0 buffer solution.

Procedure

1. Set up a pH meter and burette stand so that the pH of the sample can be measured as it is being titrated and stirred.
2. Calibrate the pH meter with pH 4.0 buffer solution and let the samples come to room temperature.
3. Pipette 50 ml of sample into a beaker.
4. Place the beaker on a magnetic stirrer (use low speed and a small bar). Record the starting pH. Also record the starting level of the HCl solution in the burette.
5. Titrate as rapidly as possible until the pH of the sample is reduced to 4.8. Record the difference between the starting level of HCl solution in

the burette and the level after titration (total amount of titrant added to the sample).

6. Perform the above steps in triplicate for each sample and average the final values.

7. Calculate alkalinity (meq l^{-1}) by the following equation, where A = the average volume of titrant (ml) and B = the normality of the acid:

$$\frac{(A)(B)(1000)}{(50)} = \text{meq alkalinity } l^{-1} \tag{26}$$

9.6   SALINITY (HYDROMETER METHOD)

The conversion of specific gravity to salinity is accomplished in two steps: (1) determination of the specific gravity with a hydrometer and correction of the figure obtained to the density in grams per millilitre at 15°C; and (2) conversion of density to salinity by using the value obtained in the first step.

Procedure

1. Fill a 500-ml graduated cylinder about two-thirds full with sample water.

2. Measure the actual specific gravity and temperature simultaneously. Estimate the specific gravity to the fourth decimal place.

3. Correct the temperature by reading from the values given in Table 9-2. For example, if the actual specific gravity is 1.022 at a temperature of 20°C, then from the table 1.0220 + 11 = 1.0231.

4. From Table 9-3, read the value for salinity. In this case, the salinity is 31.2‰

9.7   DISSOLVED OXYGEN

The procedure given here is a titration method for dissolved oxygen adapted from American Water Works Association et al. (1976). The levels of oxygen at saturation in waters of varying temperature and chlorinity are given in Table 9-4.

Sampling

Collect samples in 300-ml glass-stoppered bottles. Siphon water into the bottles with a length of airline tubing, keeping the discharge end of the

Table 9-2. Differences to Convert Hydrometer Readings at Any Temperature to Density. From Zerbe and Taylor (1953)

Observed Reading	Temperature of Water in Graduated Cylinder (°C)												
	−2.0	−1.0	0.0	1.0	2.0	3.0	4.0	5.0	6.0	7.0	8.0	9.0	10.0
0.9960													
0.9970													
0.9980													
0.9990	−1	−2	−3	−4	−5	−5	−6	−6	−6	−6	−6	−5	−5
1.0000	−2	−3	−4	−5	−5	−6	−6	−6	−6	−6	−6	−5	−5
1.0010	−3	−4	−4	−5	−6	−6	−6	−7	−7	−6	−6	−6	−5
1.0020	−3	−4	−5	−6	−6	−7	−7	−7	−7	−7	−6	−6	−5
1.0030	−4	−5	−6	−6	−7	−7	−7	−7	−7	−7	−6	−6	−5
1.0040	−4	−5	−6	−7	−7	−7	−8	−8	−7	−7	−7	−6	−6
1.0050	−5	−6	−6	−7	−8	−8	−8	−8	−8	−7	−7	−6	−6
1.0060	−6	−6	−7	−8	−8	−8	−8	−8	−8	−8	−7	−6	−6
1.0070	−6	−7	−8	−8	−8	−8	−8	−8	−8	−8	−7	−7	−6
1.0080	−7	−8	−8	−9	−9	−9	−9	−9	−8	−8	−7	−7	−6
1.0090	−7	−8	−9	−9	−9	−9	−9	−9	−9	−8	−8	−7	−6
1.0100	−8	−9	−9	−10	−10	−10	−10	−9	−9	−8	−8	−7	−6
1.0110	−9	−9	−10	−10	−10	−10	−10	−10	−9	−9	−8	−7	−6
1.0120	−9	−10	−10	−10	−10	−10	−10	−10	−10	−9	−8	−7	−7
1.0130	−10	−10	−11	−11	−11	−11	−11	−10	−10	−9	−8	−8	−7
1.0140	−10	−11	−11	−11	−11	−11	−11	−11	−10	−10	−9	−8	−7
1.0150	−11	−11	−12	−12	−12	−12	−11	−11	−10	−10	−9	−8	−7
1.0160	−12	−12	−12	−12	−12	−12	−12	−11	−11	−10	−9	−8	−7
1.0170	−12	−12	−12	−13	−13	−12	−12	−12	−11	−10	−9	−8	−7
1.0180	−13	−13	−13	−13	−13	−13	−12	−12	−11	−10	−9	−8	−7
1.0190	−13	−13	−14	−14	−13	−13	−13	−12	−12	−11	−10	−9	−8
1.0200	−14	−14	−14	−14	−14	−13	−13	−12	−12	−11	−10	−9	−8
1.0210	−14	−14	−14	−14	−14	−14	−13	−13	−12	−11	−10	−9	−8
1.0220	−15	−15	−15	−15	−15	−14	−14	−13	−12	−11	−10	−9	−8
1.0230	−15	−15	−15	−15	−15	−15	−14	−13	−12	−12	−10	−9	−8
1.0240	−16	−16	−16	−16	−15	−15	−14	−14	−13	−12	−11	−10	−8
1.0250	−16	−16	−16	−16	−16	−15	−15	−14	−13	−12	−11	−10	−8
1.0260	−17	−17	−17	−16	−16	−16	−15	−14	−13	−12	−11	−10	−8
1.0270	−18	−17	−17	−17	−17	−16	−15	−14	−14	−12	−11	−10	−9
1.0280	−18	−18	−18	−17	−.17	−16	−16	−15	−14	−13	−11	−10	−9
1.0290	−19	−18	−18	−18	−17	−17	−16	−15	−14	−13	−12	−10	−9
1.0300	−19	−19	−19	−18	−18	−17	−16	−15	−14	−13	−12	−10	−9
1.0310	−20	−19	−19	−19	−18	−17	−16	−16	−15	−13	−12	−10	−9

Table 9-2.   (Continued)

Observed Reading	Temperature of Water in Graduated Cylinder (°C)											
	11.0	12.0	13.0	14.0	15.0	16.0	17.0	18.0	18.5	19.0	19.5	20.0
0.9960												
0.9970												
0.9980							3	4	5	6	7	8
0.9990	−4	−3	−2	−1	0	1	3	4	5	6	7	8
1.0000	−4	−3	−2	−1	0	1	3	4	5	6	7	8
1.0010	−4	−3	−2	−1	0	1	3	4	5	6	7	8
1.0020	−4	−3	−2	−1	0	1	3	4	5	6	7	8
1.0030	−4	−3	−2	−1	0	1	3	4	5	6	7	8
1.0040	−5	−4	−3	−1	0	2	3	5	6	6	7	8
1.0050	−5	−4	−3	−1	0	2	3	5	6	7	8	9
1.0060	−5	−4	−3	−1	0	2	3	5	6	7	8	9
1.0070	−5	−4	−3	−2	0	2	3	5	6	7	8	9
1.0080	−5	−4	−3	−2	0	2	3	5	6	7	8	9
1.0090	−5	−4	−3	−2	0	2	3	5	6	7	8	9
1.0100	−5	−4	−3	−2	0	2	3	5	6	7	8	9
1.0110	−5	−4	−3	−2	0	2	3	5	6	7	8	9
1.0120	−6	−4	−3	−2	0	2	3	5	6	7	8	9
1.0130	−6	−4	−3	−2	0	2	4	5	6	7	8	10
1.0140	−6	−4	−3	−2	0	2	4	5	6	8	9	10
1.0150	−6	−4	−3	−2	0	2	4	5	6	8	9	10
1.0160	−6	−5	−3	−2	0	2	4	6	7	8	9	10
1.0170	−6	−5	−3	−2	0	2	4	6	7	8	9	10
1.0180	−6	−5	−3	−2	0	2	4	6	7	8	9	10
1.0190	−6	−5	−3	−2	0	2	4	6	7	8	9	10
1.0200	−6	−5	−3	−2	0	2	4	6	7	8	9	10
1.0210	−6	−5	−3	−2	0	2	4	6	7	8	9	10
1.0220	−7	−5	−3	−2	0	2	4	6	7	8	9	11
1.0230	−7	−5	−4	−2	0	2	4	6	7	8	9	11
1.0240	−7	−5	−4	−2	0	2	4	6	7	8	10	11
1.0250	−7	−5	−4	−2	0	2	4	6	7	8	10	11
1.0260	−7	−5	−4	−2	0	2	4	6	7	9	10	11
1.0270	−7	−5	−4	−2	0	2	4	6	7	9	10	11
1.0280	−7	−6	−4	−2	0	2	4	6	8	9	10	11
1.0290	−7	−6	−4	−2	0	2	4	6	8	9	10	11
1.0300	−7	−6	−4	−2	0	2	4	6	8	9	10	12
1.0310	−8	−6	−4	−2	0	2	4					

Table 9-2.   (Continued)

Observed Reading	\multicolumn{13}{Temperature of Water in Graduated Cylinder (°C)}

Observed Reading	20.5	21.0	21.5	22.0	22.5	23.0	23.5	24.0	24.5	25.0	25.5	26.0	26.5
0.9960											19	20	21
0.9970			10	11	12	14	15	16	17	18	19	20	22
0.9980	9	10	11	12	13	14	15	16	17	18	19	21	22
0.9990	9	10	11	12	13	14	15	16	17	18	20	21	22
1.0000	9	10	11	12	13	14	15	16	17	19	20	21	22
1.0010	9	10	11	12	13	14	15	17	18	19	20	21	23
1.0020	9	10	11	12	13	14	16	17	18	19	20	22	23
1.0030	9	10	11	12	13	15	16	17	18	19	21	22	23
1.0040	9	10	11	12	14	15	16	17	18	20	21	22	23
1.0050	10	11	12	13	14	15	16	17	19	20	21	22	24
1.0060	10	11	12	13	14	15	16	18	19	20	21	23	24
1.0070	10	11	12	13	14	15	17	18	19	20	21	23	24
1.0080	10	11	12	13	14	16	17	18	19	20	22	23	24
1.0090	10	11	12	13	15	16	17	18	19	21	22	23	25
1.0100	10	11	12	14	15	16	17	18	20	21	22	24	25
1.0110	10	12	13	14	15	16	17	19	20	21	22	24	25
1.0120	10	12	13	14	15	16	18	19	20	21	23	24	25
1.0130	11	12	13	14	15	16	18	19	20	22	23	24	26
1.0140	11	12	13	14	15	17	18	19	20	22	23	24	26
1.0150	11	12	13	14	16	17	18	20	21	22	23	25	26
1.0160	11	12	13	14	16	17	18	20	21	22	24	25	26
1.0170	11	12	13	15	16	17	18	20	21	22	24	25	27
1.0180	11	12	14	15	16	17	19	20	21	23	24	25	27
1.0190	11	12	14	15	16	18	19	20	21	23	24	26	27
1.0200	11	13	14	15	16	18	19	20	22	23	24	26	27
1.0210	12	13	14	15	17	18	19	21	22	23	25	26	27
1.0220	12	13	14	15	17	18	19	21	22	23	25	26	28
1.0230	12	13	14	16	17	18	20	21	22	24	25	26	28
1.0240	12	13	14	16	17	18	20	21	22	24	25	27	28
1.0250	12	13	15	16	17	18	20	21	23	24	25	27	28
1.0260	12	13	15	16	17	19	20	22	23	24	26	27	29
1.0270	12	14	15	16	17	19	20	22	23	24	26	27	29
1.0280	12	14	15	16	18	19	20	22	23	25	26	28	29
1.0290	13	14	15	16	18	19	21	22	23				
1.0300	13	14	15	16	18								
1.0310													

Table 9-2. (Continued)

Observed Reading	Temperature of Water in Graduated Cylinder (°C)												
	27.0	27.5	28.0	28.5	29.0	29.5	30.0	30.5	31.0	31.5	32.0	32.5	33.0
0.9960	23	24	25	27	28	29	31	32	34	35	37	38	40
0.9970	23	24	26	27	28	30	31	33	34	36	37	39	40
0.9980	23	25	26	27	29	30	31	33	34	36	38	39	41
0.9990	24	25	26	28	29	30	32	33	35	36	38	39	41
1.0000	24	25	26	28	29	31	32	34	35	37	38	40	41
1.0010	24	25	27	28	30	31	32	34	35	37	39	40	42
1.0020	24	26	27	28	30	31	33	34	36	37	39	41	42
1.0030	25	26	27	29	30	32	33	35	36	38	39	41	42
1.0040	25	26	28	29	30	32	33	35	36	38	40	41	43
1.0050	25	26	28	29	31	32	34	35	37	38	40	42	43
1.0060	25	27	28	30	31	32	34	36	37	39	40	42	44
1.0070	26	27	28	30	31	33	34	36	38	39	41	42	44
1.0080	26	27	29	30	32	33	35	36	38	39	41	43	44
1.0090	26	28	29	30	32	33	35	36	38	40	41	43	45
1.0100	26	28	29	31	32	34	35	37	38	40	42	43	45
1.0110	27	28	30	31	32	34	36	37	39	40	42	44	45
1.0120	27	28	30	31	33	34	36	37	39	41	42	44	46
1.0130	27	29	30	32	33	35	36	38	39	41	43	44	46
1.0140	27	29	30	32	33	35	36	38	40	41	43	45	46
1.0150	28	29	31	32	34	35	37	38	40	42	43	45	47
1.0160	28	29	31	32	34	35	37	39	40	42	44	45	47
1.0170	28	30	31	33	34	36	37	39	40	42	44	46	47
1.0180	28	30	31	33	34	36	38	39	41	42	44	46	48
1.0190	29	30	32	33	35	36	38	39	41	43	44	46	48
1.0200	29	30	32	33	35	37	38	40	41	43	45	47	48
1.0210	29	31	32	34	35	37	38	40	42	43	45	47	49
1.0220	29	31	32	34	36	37	39	40	42	44	45	47	49
1.0230	30	31	33	34	36	37	39	41	42	44	46	47	49
1.0240	30	31	33	34	36	37	39	41	42	44	46	48	49
1.0250	30	31	33	35	36	38	39	41	43	44	46	48	50
1.0260	30	32	33	35	37	38	40	41	43	45	46	48	50
1.0270	30	32	34	35	37	38	40						
1.0280	31	32											
1.0290													
1.0300													
1.0310													

Table 9-3. Corresponding Densities and Salinities.* From Zerbe and Taylor (1953)

Density	Salinity	Density	Salinity	Density	Salinity	Density	Salinity
0.9991	0.0	1.0036	5.8	1.0081	11.6	1.0126	17.5
0.9992	0.0	1.0037	5.9	1.0082	11.8	1.0127	17.7
0.9993	0.2	1.0038	6.0	1.0083	11.9	1.0128	17.8
0.9994	0.3	1.0039	6.2	1.0084	12.0	1.0129	17.9
0.9995	0.4	1.0040	6.3	1.0085	12.2	1.0130	18.0
0.9996	0.6	1.0041	6.4	1.0086	12.3	1.0131	18.2
0.9997	0.7	1.0042	6.6	1.0087	12.4	1.0132	18.3
0.9998	0.8	1.0043	6.7	1.0088	12.6	1.0133	18.4
0.9999	0.9	1.0044	6.8	1.0089	12.7	1.0134	18.6
1.0000	1.1	1.0045	6.9	1.0090	12.8	1.0135	18.7
1.0001	1.2	1.0046	7.1	1.0091	12.9	1.0136	18.8
1.0002	1.3	1.0047	7.2	1.0092	13.1	1.0137	19.0
1.0003	1.5	1.0048	7.3	1.0093	13.2	1.0138	19.1
1.0004	1.6	1.0049	7.5	1.0094	13.3	1.0139	19.2
1.0005	1.7	1.0050	7.6	1.0095	13.5	1.0140	19.3
1.0006	1.9	1.0051	7.7	1.0096	13.6	1.0141	19.5
1.0007	2.0	1.0052	7.9	1.0097	13.7	1.0142	19.6
1.0008	2.1	1.0053	8.0	1.0098	13.9	1.0143	19.7
1.0009	2.2	1.0054	8.1	1.0099	14.0	1.0144	19.9
1.0010	2.4	1.0055	8.2	1.0100	14.1	1.0145	20.0
1.0011	2.5	1.0056	8.4	1.0101	14.2	1.0146	20.1
1.0012	2.6	1.0057	8.5	1.0102	14.4	1.0147	20.3
1.0013	2.8	1.0058	8.6	1.0103	14.5	1.0148	20.4
1.0014	2.9	1.0059	8.8	1.0104	14.6	1.0149	20.5
1.0015	3.0	1.0060	8.9	1.0105	14.8	1.0150	20.6
1.0016	3.2	1.0061	9.0	1.0106	14.9	1.0151	20.8
1.0017	3.3	1.0062	9.2	1.0107	15.0	1.0152	20.9
1.0018	3.4	1.0063	9.3	1.0108	15.2	1.0153	21.0
1.0019	3.5	1.0064	9.4	1.0109	15.3	1.0154	21.2
1.0020	3.7	1.0065	9.6	1.0110	15.4	1.0155	21.3
1.0021	3.8	1.0066	9.7	1.0111	15.6	1.0156	21.4
1.0022	3.9	1.0067	9.8	1.0112	15.7	1.0157	21.6
1.0023	4.1	1.0068	9.9	1.0113	15.8	1.0158	21.7
1.0024	4.2	1.0069	10.1	1.0114	16.0	1.0159	21.8
1.0025	4.3	1.0070	10.2	1.0115	16.1	1.0160	22.0
1.0026	4.5	1.0071	10.3	1.0116	16.2	1.0161	22.1
1.0027	4.6	1.0072	10.5	1.0117	16.3	1.0162	22.2
1.0028	4.7	1.0073	10.6	1.0118	16.5	1.0163	22.4
1.0029	4.8	1.0074	10.7	1.0119	16.6	1.0164	22.5
1.0030	5.0	1.0075	10.8	1.0120	16.7	1.0165	22.6

*Density at 15°C. Salinity in parts per thousand.

Table 9-3. (Continued)

Density	Salinity	Density	Salinity	Density	Salinity	Density	Salinity
1.0031	5.1	1.0076	11.0	1.0121	16.9	1.0166	22.7
1.0032	5.2	1.0077	11.1	1.0122	17.0	1.0167	22.9
1.0033	5.4	1.0078	11.2	1.0123	17.1	1.0168	23.0
1.0034	5.5	1.0079	11.4	1.0124	17.3	1.0169	23.1
1.0035	5.6	1.0080	11.5	1.0125	17.4	1.0170	23.3
1.0171	23.4	1.0211	28.6	1.0251	33.8	1.0291	39.0
1.0172	23.5	1.0212	28.8	1.0252	34.0	1.0292	39.2
1.0173	23.7	1.0213	28.9	1.0253	34.1	1.0293	39.3
1.0174	23.8	1.0214	29.0	1.0254	34.2	1.0294	39.4
1.0175	23.9	1.0215	29.1	1.0255	34.4	1.0295	39.6
1.0176	24.1	1.0216	29.3	1.0256	34.5	1.0296	39.7
1.0177	24.2	1.0217	29.4	1.0257	34.6	1.0297	39.8
1.0178	24.3	1.0218	29.5	1.0258	34.8	1.0298	39.9
1.0179	24.4	1.0219	29.7	1.0259	34.9	1.0299	40.1
1.0180	24.6	1.0220	29.8	1.0260	35.0	1.0300	40.2
1.0181	24.7	1.0221	29.9	1.0261	35.1	1.0301	40.3
1.0182	24.8	1.0222	30.1	1.0262	35.3	1.0302	40.4
1.0183	25.0	1.0223	30.2	1.0263	35.4	1.0303	40.6
1.0184	25.1	1.0224	30.3	1.0264	35.5	1.0304	40.7
1.0185	25.2	1.0225	30.4	1.0265	35.7	1.0305	40.8
1.0186	25.4	1.0226	30.6	1.0266	35.8	1.0306	41.0
1.0187	25.5	1.0227	30.7	1.0267	35.9	1.0307	41.1
1.0188	25.6	1.0228	30.8	1.0268	36.0	1.0308	41.2
1.0189	25.8	1.0229	31.0	1.0269	36.2	1.0309	41.4
1.0190	25.9	1.0230	31.1	1.0270	36.3	1.0310	41.5
1.0191	26.0	1.0231	31.2	1.0271	36.4	1.0311	41.6
1.0192	26.1	1.0232	31.4	1.0272	36.6	1.0312	41.7
1.0193	26.3	1.0233	31.5	1.0273	36.7	1.0313	41.9
1.0194	26.4	1.0234	31.6	1.0274	36.8	1.0314	42.0
1.0195	26.5	1.0235	31.8	1.0275	37.0	1.0315	42.1
1.0196	26.7	1.0236	31.9	1.0276	37.1	1.0316	42.3
1.0197	26.8	1.0237	32.0	1.0277	37.2	1.0317	42.4
1.0198	26.9	1.0238	32.1	1.0278	37.3	1.0318	42.5
1.0199	27.1	1.0239	32.3	1.0279	37.5	1.0319	42.7
1.0200	27.2	1.0240	32.4	1.0280	37.6	1.0320	42.8
1.0201	27.3	1.0241	32.5	1.0281	37.7		
1.0202	27.5	1.0242	32.7	1.0282	37.9		
1.0203	27.6	1.0243	32.8	1.0283	38.0		
1.0204	27.7	1.0244	32.9	1.0284	38.1		
1.0205	27.8	1.0245	33.1	1.0285	38.2		

Table 9-3.   (Continued)

Density	Salinity	Density	Salinity	Density	Salinity	Density	Salinity
1.0206	28.0	1.0246	33.2	1.0286	38.4		
1.0207	28.1	1.0247	33.3	1.0287	38.5		
1.0208	28.2	1.0248	33.5	1.0288	38.6		
1.0209	28.4	1.0249	33.6	1.0289	38.8		
1.0210	28.5	1.0250	33.7	1.0290	38.9		

Table 9-4.   Dissolved Oxygen (mg O_2 l^{-1}) at Saturation in Freshwater, Brackish Water, and Seawater at Different Temperatures. Calculated from Data in Murray and Riley (1969)

Temperature (°C)	Chlorinity (‰)										
	0	2	4	6	8	10	12	14	16	18	20
1	14.24	13.87	13.54	13.22	12.91	12.59	12.29	11.99	11.70	11.42	11.15
2	13.84	13.50	13.18	12.88	12.56	12.26	11.98	11.69	11.40	11.13	10.86
3	13.45	13.14	12.84	12.55	12.25	11.96	11.68	11.39	11.12	10.85	10.59
4	13.09	12.79	12.51	12.22	11.93	11.65	11.38	11.10	10.83	10.59	10.34
5	12.75	12.45	12.18	11.91	11.63	11.36	11.09	10.83	10.57	10.33	10.10
6	12.44	12.15	11.86	11.60	11.33	11.07	10.82	10.56	10.32	10.09	9.86
7	12.13	11.85	11.58	11.32	11.06	10.82	10.56	10.32	10.07	9.84	9.63
8	11.85	11.56	11.29	11.05	10.80	10.56	10.32	10.07	9.84	9.61	9.40
9	11.56	11.29	11.02	10.77	10.54	10.30	10.07	9.84	9.61	9.40	9.20
10	11.29	11.03	10.77	10.53	10.30	10.07	9.84	9.61	9.40	9.20	9.00
11	11.05	10.77	10.53	10.29	10.07	9.84	9.63	9.41	9.20	9.00	8.80
12	10.80	10.53	10.29	10.06	9.84	9.63	9.41	9.21	9.00	8.80	8.61
13	10.56	10.30	10.07	9.84	9.63	9.41	9.21	9.01	8.81	8.61	8.42
14	10.33	10.07	9.86	9.63	9.41	9.21	9.01	8.81	8.62	8.44	8.25
15	10.10	9.86	9.64	9.43	9.23	9.03	8.83	8.64	8.44	8.27	8.09
16	9.89	9.66	9.44	9.24	9.03	8.84	8.64	8.47	8.28	8.11	7.94
17	9.67	9.46	9.26	9.05	8.85	8.65	8.47	8.30	8.11	7.94	7.78
18	9.47	9.27	9.07	8.87	8.67	8.48	8.31	8.14	7.97	7.79	7.64
19	9.28	9.08	8.88	8.68	8.50	8.31	8.15	7.98	7.08	7.65	7.49
20	9.11	8.90	8.70	8.51	8.32	8.15	7.99	7.84	7.66	7.51	7.36
21	8.93	8.72	8.54	8.35	8.17	7.99	7.84	7.69	7.52	7.38	7.23
22	8.75	8.55	8.38	8.19	8.02	7.85	7.69	7.54	7.39	7.25	7.11
23	8.60	8.40	8.22	8.04	7.87	7.71	7.55	7.41	7.26	7.12	6.99
24	8.44	8.25	8.07	7.89	7.72	7.56	7.42	7.28	7.13	6.99	6.86
25	8.27	8.09	7.92	7.75	7.58	7.44	7.29	7.15	7.01	6.88	6.75
26	8.12	7.94	7.78	7.62	7.45	7.31	7.16	7.03	6.89	6.76	6.63
27	7.98	7.79	7.64	7.49	7.32	7.18	7.03	6.91	6.78	6.65	6.52
28	7.84	7.65	7.51	7.36	7.19	7.06	6.92	6.79	6.66	6.53	6.40
29	7.69	7.52	7.38	7.23	7.08	6.95	6.82	6.68	6.55	6.42	6.29
30	7.56	7.39	7.25	7.12	6.96	6.83	6.70	6.58	6.45	6.32	6.19

siphon tube completely submerged. Allow water from the aquarium to overflow the bottle three to four times its volume. Insert the stopper without trapping air. Analyze within 2 hr.

Reagents

Manganous sulfate solution.   Dissolve any of the following manganous sulfates (hydrous) in a little distilled water, then filter and dilute to 100 ml: 48.0 g $MnSO_4·4H_2O$, 40.0 g $MnSO_4·2H_2O$, or 36.4 g $MnSO_4·H_2O$.

Alkali-iodide-azide solution.   Dissolve 50.0 g NaOH (or 70.0 g KOH) and 15.0 g KI in distilled water and dilute to 100 ml. Add 1.0 g NaN_3 dissolved in 4.0 ml of distilled water.

Starch solution.   Blend 0.5–0.6 g of soluble starch in a beaker with a little distilled water. Add to a beaker containing 100 ml of boiling water and continue boiling for a few minutes. Add a few drops of chloroform to preserve.

Sodium thiosulfate standard solution.   Dissolve 2.48 g $Na_2S_2O_3·5H_2O$ in boiled and cooled distilled water. Dilute to 100 ml with distilled water and add 0.5 ml of chloroform to preserve.

Standard sodium thiosulfate titrant (0.025 N).   Dilute 25.0 ml of sodium thiosulfate stock solution to 100 ml with distilled water. Add 0.5 ml of chloroform to preserve. 1.0 ml of titrant = 0.20 mg O_2 1.00 ml^{-1}.

Sulfuric acid.   Concentrated (36.0 N).

Potassium dichromate standard solution (0.025 N).   Dry some $K_2Cr_2O_7$ for 2 hr in a desiccator at 103°C. Weigh 0.0225 g and dissolve in 100 ml of distilled water in a volumetric flask.

Standardization

1. Dissolve 2.0 g KI in 100–150 ml of distilled water. Add 10.0 ml of a solution containing 1 part sulfuric acid and 9 parts distilled water.

2. Add 20.0 ml of potassium dichromate standard solution.

3. Dilute to 400 ml with distilled water and titrate with the thiosulfate titrant. Add starch (1–2 ml) near the end point of the titration when a pale yellow color is reached. 20.0 ml of titrant is required when the thiosulfate titrant solution is exactly 0.025 N.

Procedure

1. Add 2.0 ml of manganous sulfate solution, followed by 2.0 ml of alkali-iodide-azide solution. Keep the end of the pipette submerged in the water sample.

2. Replace the stopper without trapping air. Invert the bottle several times, allowing the precipitate to settle halfway each time before inverting again. *Note*. In brackish water and seawater samples, wait 10 min before proceeding to the next step.

3. Add 2.0 ml sulfuric acid to the sample by letting it run out of the pipette down the inside of the bottle neck. Restopper and invert gently several times until the precipitate dissolves.

4. Pour 200 ml of sample into a beaker. Titrate with sodium thiosulfate titrant to a pale yellow color. Add 1–2 ml of starch solution and continue titration until the pale blue color disappears. After the starch has been added, avoid going past the end point by adding the titrant a drop at a time and gently swirling the beaker after each drop. 1.0 ml of titrant = 1.0 mg O_2 l^{-1}. *Note*. If the amount of titrant used during standardization is not exactly 20.0 ml (step 2), then at the end of the test procedure the amount of dissolved oxygen in the sample can be determined by

$$mg\ O_2\ l^{-1} = \frac{\text{ml test titrant} \times 10}{\text{actual ml standardization titrant}} \tag{27}$$

Literature Cited

American Public Health Association, American Water Works Association, and Water Pollution Control Federation.
  1976   Standard methods for the examination of water and wastewater, 14th ed. Am. Pub. Health Assoc., Washington, D.C., 1193 pp.

Amlacher, E., D. A. Conroy, and R. L. Herman.
  1970   Textbook of fish diseases. TFH Publ., Neptune City, N.J., 302 pp.

Anderson, D. P.
  1974   Diseases of fishes: fish immunology. TFH Publ., Neptune City, N.J., 239 pp.

Anikouchine, W. A. and R. W. Sternberg.
  1973   The world ocean: an introduction to oceanography. Prentice-Hall, Englewood Cliffs, N.J., 338 pp.

Anonymous.
  1971   Starting with sea water. II. Aeration and filtration. *Petfish Mon.* **6:** 21–22.

Anthonisen, A. C., R. C. Loehr, T. B. S. Prakasam, and E. G. Srinath.
  1976   Inhibition of nitrification by ammonia and nitrous acid. *J. Water Pollut. Control Fed.* **48:** 835–852.

Armstrong, D. A., D. Chippendale, A. W. Knight, and J. E. Colt.
  1978   Interaction of ionized and un-ionized ammonia on short-term survival and growth of prawn larvae, *Macrobrachium rosenbergii. Biol. Bull.* **154:** 15–31.

Armstrong, D. A., M. J. Stephenson, and A. W. Knight.
  1976   Acute toxicity of nitrite to larvae of the giant Malaysian prawn, *Macrobrachium rosenbergii. Aquaculture* **9:** 39–46.

Arnon, D. I. and P. R. Stout.
  1939   The essentiality of certain elements in minute quantity for plants, with special reference to copper. *Plant Physiol.* **14:** 371–375.

Baines, G. W.
  1975   Blood pH effects in eight fishes from the teleostean family Scorpaenidae. *Comp. Biochem. Physiol.* **51A:** 833–843.

Baker, J. T. P.
  1969   Histological and electron microscopical observations on copper poisoning in the winter flounder (*Pseudopleuronectes americanus*). *J. Fish. Res. Board Can.* **26:** 2785–2793.

Barber, R. T.
  1966   Interaction of bubbles and bacteria in the formation of organic aggregates in sea water. *Nature* **211:** 257–258.

Barcelona, M. J., T. R. Tosteson, and D. K. Atwood.
  1976   Study of organic-calcium interactions: gypsum precipitation in tropical surface waters. *Mar. Chem.* **4:** 89–92.

Barica, J.
  1973   Reliability of an ammonia probe for electrometric determination of total ammonia nitrogen in fish tanks. *J. Fish. Res. Board Can.* **30:** 1389–1392.

Basu, S. P.
  1959   Active respiration of fish in relation to ambient concentrations of oxygen and carbon dioxide. *J. Fish. Res. Board Can.* **16:** 175–212.

Batoosingh, E., G. A. Riley, and B. Keshwar.
  1969   An analysis of experimental methods for producing particulate organic matter in sea water by bubbling. *Deep-Sea Res.* **16:** 213–219.

Battelle Memorial Institute.
  1969   Ammonia removal from agricultural runoff and secondary effluents by selected ion exchange. Report No. TWRC-5, Fed. Water Pollut. Control Adm., Cincinnati, 52 pp.

Baylor, E. R., W. H. Sutcliffe, and D. S. Hirschfield.
  1962   Adsorption of phosphate onto bubbles. *Deep-Sea Res.* **9:** 120–124.

Berner, R. A.
  1968   Calcium carbonate concretions formed by the decomposition of organic matter. *Science* **159:** 195–197.

Berner, R. A.
  1975   The role of magnesium in the crystal growth of calcite and aragonite from sea water. *Geochim. Cosmochim. Acta.* **39:** 489–504.

Black, E. C., F. E. J. Fry, and V. S. Black.
  1954   The influence of carbon dioxide on the utilization of oxygen by some freshwater fish. *Can. J. Zool.* **32:** 408–420.

Bowen, H. J. M.
  1966   Trace elements in biochemistry. Academic Press, New York, 241 pp.

Bower, C. E. and J. P. Bidwell.
  1978   Ionization of ammonia in seawater: effects of temperature, pH, and salinity. *J. Fish. Res. Board Can.* **35:** 1012–1016.

Breder, C. M. Jr. and H. W. Smith.
  1932   On use of sodium bicarbonate and calcium in the rectification of sea-water aquaria. *J. Mar. Biol. Assoc. N. S.* **18:** 199–200.

Brett, J. R.
  1956   Some principles in the thermal requirements of fishes. *Quart. Rev. Biol.* **31:** 75–87.

Brockway, D. R.
  1950   Metabolic products and their effects. *Prog. Fish-Cult.* **12:** 127–129.

Brown, D. A. and D. J. McLeay.
  1975   Effect of nitrite on methemoglobin and total hemoglobin of juvenile rainbow trout. *Prog. Fish-Cult.* **37:** 36–38.

Buelow, R. W., K. L. Kropp, J. Withered, and J. M. Symons.
  1975   Nitrate removal by anion-exchange resins. *J. Am. Water Works Assoc.* **67:** 528–534.

Burkhalter, D. E. and C. M. Kaya.
  1977   Effects of prolonged exposure to ammonia on fertilized eggs and sac fry of rainbow trout (*Salmo gairdneri*). *Trans. Amer. Fish. Soc.* **106:** 470–475.

Burton, D. T., A. H. Jones, and J. Cairns Jr.
  1972 Acute zinc toxicity to rainbow trout (*Salmo gairdneri*): confirmation of the hypothesis that death is related to tissue hypoxia. *J. Fish. Res. Board Can.* **29:** 1463–1466.

Bullock, G. L.
  1971 Diseases of fishes, book 2B: the identification of fish pathogenic bacteria. TFH Publ., Neptune City, N.J., 41 pp.

Bullock, G. L., D. A. Conroy, and S. F. Snieszko.
  1971 Diseases of fishes, book 2A: bacterial diseases of fishes. TFH Publ., Neptune City, N.J., 151 pp.

Bullock, G. L. and H. M. Stuckey.
  1977 Ultraviolet treatment of water for destruction of five gram-negative bacteria pathogenic to fishes. *J. Fish. Res. Board Can.* **34:** 1244–1249.

Cardeilhac, P. T. and E. R. Hall.
  1977 Acute copper poisoning of cultured marine teleosts. *Am. J. Vet. Res.* **38:** 525–527.

Castro, W. E., P. B. Zielinski, and P. A. Sandifer.
  1975 Performance characteristics of airlift pumps of short length and small diameter. *Proc. 6th Ann. Meet. World Maricult Soc.*, J. W. Avault Jr. and R. Miller (eds.). World Maricult. Soc., La. State Univ., Baton Rouge, pp. 451-461.

Chave, K. E.
  1965 Calcium carbonate: association with organic matter in surface seawater. *Science* **148:** 1723–1724.

Chave, K. E., K. S. Deffeyes, P. K. Weyl, R. M. Garrels, and M. E. Thompson.
  1962 Observations on the solubility of skeletal carbonates in aqueous solutions. *Science* **137:** 33–34.

Chave, K. E. and E. Suess.
  1967 Suspended minerals in seawater. *Trans. N. Y. Acad. Sci. (Ser. II)* **29:** 991–1000.

Chave, K. E. and E. Suess.
  1970 Calcium carbonate saturation in seawater: effects of dissolved organic matter. *Limnol. Oceanogr.* **15:** 633–637.

Christensen, G. M.
  1971/1972 Effects of metal cations and other chemicals upon the *in vitro* activity of two enzymes in the blood plasma of the white sucker, *Catostomus commersoni* (Lacepede). *Chem.-Biol. Interactions* **4:** 351–361.

Collins, M. T., J. B. Gratzek, D. L. Dawe, and T. G. Nemetz.
  1975 Effects of parasiticides on nitrification. *J. Fish. Res. Board Can.* **32:** 2033–2037.

Collins, M. T., J. B. Gratzek, D. L. Dawe, and T. G. Nemetz.
  1976 Effects of antibacterial agents on nitrification in an aquatic recirculating system. *J. Fish. Res. Board Can.* **33:** 215–218.

Conrad, J. F., R. A. Holt, and T. D. Kreps.
  1975 Ozone disinfection of flowing water. *Prog. Fish-Cult.* **37:** 134–135.

Conroy, D. A.
  1968 Partial bibliography on the bacterial diseases of fish: an annotated bibliography for the years 1870-1966. FAO Fish. Tech. Pap. No. 73, Food Agric. Organ. Organ., U. N., Rome, 75 pp.

Cooper, L. H. N.
    1932   On the effect of long continued additions of lime to aquarium sea-water. *J. Mar. Biol. Assoc. N. S.* **18:** 201–202.

Crawford, R. E. and G. H. Allen.
    1977   Seawater inhibition of nitrite toxicity to chinook salmon. *Trans. Am. Fish. Soc.* **106:** 105–109.

Crawshaw, L. I.
    1977   Physiological and behavioral reactions of fishes to temperature change. *J. Fish. Res. Board Can.* **34:** 730–734.

Craft, T. F.
    1966   Review of rapid sand filtration theory. *J. Am. Water Works Assoc.* **58:** 428–439.

Culp, G. and A. Slechta.
    1966   Nitrogen removal from waste effluents. *Public Works* **97:** 90–91.

Davey, E. W., J. H. Gentile, S. J. Erickson, and P. Betzer.
    1970   Removal of trace metals from marine culture media. Pap. presented at 33rd Ann. Meet., Am. Soc. Limnol. Oceanogr., Univ. R. I., Providence.

Dejours, P., A. Toulmond, and J. P. Truchot.
    1977   The effect of hyperoxia on the breathing of marine fishes. *Comp. Biochem. Physiol.* **58A:** 409–411.

Doudoroff, P. and M. Katz.
    1953   Critical review of literature on the toxicity of industrial wastes and their components to fish. II. The metals as salts. *Sewage Ind. Wastes* **25:** 802–839.

Downing, K. M. and J. C. Merkens.
    1955   The influence of dissolved-oxygen concentration on the toxicity of un-ionized ammonia to rainbow trout (*Salmo gairdneri* Richardson). *Ann. Appl. Biol.* **43:** 243–246.

Duijn, C. van Jr.
    1973   Diseases of fish, 3rd ed. Iliffe Books, London, 372 pp.

Dulin, M. P.
    1976   Diseases of marine aquarium fishes. TFH Publ., Neptune City, N.J., 128 pp.

Eddy, F. B., J. P. Lomholt, R. E. Weber, and K. Johansen.
    1977   Blood respiratory properties of rainbow trout (*Salmo gairdneri*) kept in water of high CO_2 tension. *J. Exp. Biol.* **67:** 37–48.

Eliassen, R., B. M. Wyckoff, and C. D. Tonkin.
    1965   Ion exchange for reclamation of reusable supplies. *J. Am. Water Works Assoc.* **57:** 1113–1122.

Ellis, M. M.
    1937   Detection and measurement of stream pollution. *Bull. Bur. Fish.* **48:** 365–437.

Emerson, K., R. C. Russo, R. E. Lund, and R. V. Thurston.
    1975   Aqueous ammonia equilibrium calculations: effect of pH and temperature. *J. Fish. Res. Board Can.* **32:** 2379–2383.

Epifanio, C. E. and R. F. Srna.
    1975   Toxicity of ammonia, nitrite ion, nitrate ion, and orthophosphate to *Mercenaria mercenaria* and *Crassostrea virginica. Mar. Biol.* **33:** 241–246.

Evans, D. H.
    1973   Sodium uptake by the sailfin molly, *Poecilia latipinna:* kinetic analysis of a

carrier system present in both freshwater-acclimated and seawater-acclimated individuals. *Comp. Biochem. Physiol.* **45A:** 843–850.

Evans, D. H.
1975 The effects of various external cations and sodium transport inhibitors on sodium uptake by the sailfin molly, *Poecilia latipinna,* acclimated to sea water. *J. Comp. Physiol.* **96B:** 111–115.

Farooq, S., E. S. K. Chian, and R. S. Englebrecht.
1977a Basic concepts in disinfection with ozone. *J. Water Pollut. Control Fed.* **49:** 1818–1831.

Farooq, S., R. S. Englebrecht, and E. S. K. Chian.
1977b Influence of temperature and U. V. light on disinfection with ozone. *Water Res.* **11:** 737–741.

Fitzgerald, G. P.
1963 Factors affecting the toxicity of copper to algae and fish. *Proc. Am. Chem. Soc. Meet.,* New York, pp. 21–24.

Folk, R. L.
1974 The natural history of crystalline calcium carbonate: effect of magnesium content and salinity. *J. Sediment. Petr.* **44:** 40–53.

Frisch, N. W. and R. Kunin.
1960 Organic fouling of anion-exchange resins. *J. Am. Water Works Assoc.* **52:** 875–887.

Fromm, P. O. and J. R. Gillette.
1968 Effect of ambient ammonia on blood ammonia and nitrogen excretion of rainbow trout (*Salmo gairdneri*). *Comp. Biochem. Physiol.* **26:** 887–896.

Fry, F. E. J.
1947 Effects of environment on animal activity. *Publ. Ont. Fish. Res. Lab.* **68:** 1–62.

Fry, F. E. J. and J. S. Hart.
1948 The relation of temperature to oxygen consumption in the goldfish. *Biol. Bull.* **94:** 66–77.

Gardner, G. R. and G. LaRoche.
1973 Copper induced lesions in estuarine teleosts. *J. Fish. Res. Board Can.* **30:** 363–368.

Gilbert, T. R. and A. M. Clay.
1973 Determination of ammonia in aquaria and in sea water using the ammonia electrode. *Anal. Chem.* **45:** 1758–1759.

Grasshoff, K. (ed.).
1976 Methods of seawater analysis. Verlag Chemie, Weinheim, 317 pp.

Green, A. A. and R. W. Root.
1933 The equilibrium between hemoglobin and oxygen in the blood of certain fishes. *Biol. Bull.* **64:** 383–404.

Gundersen, K. and C. W. Mountain.
1973 Oxygen utilization and pH changes in the ocean resulting from biological nitrate formation. *Deep-Sea Res.* **20:** 1083–1091.

Hall, F. G. and F. H. McCutcheon.
1938 The affinity of hemoglobin for oxygen in marine fishes. *J. Cell. Comp. Physiol.* **11:** 205–212.

Hampson, B. L.
    1976   Ammonia concentration in relation to ammonia toxicity during a rainbow trout rearing experiment in a closed freshwater-seawater system. *Aquaculture* **9:** 61–70.

Hassen-Teufel, W., R. Jagitsch, and F. F. Koczy.
    1963   Impregnation of glass surface against sorption of phosphate traces. *Limnol. Oceanogr.* **8:** 152–156.

Herald, E. S., R. P. Dempster, and M. Hunt.
    1970   Ultraviolet sterilization of aquarium water. *Drum and Croaker* (Spec. ed.), U. S. Dept. Inter., W. Hagen (ed.), pp. 57–71.

Heron, J.
    1962   Determination of phosphate in water after storage in polyethylene. *Limnol. Oceanogr.* **7:** 316–321.

Hirayama, K.
    1965   Studies on water control by filtration through sand bed in a marine aquarium with closed circulating system. I. Oxygen consumption during filtration as an index in evaluating the degree of purification of breeding water. *Bull. Jap. Soc. Sci. Fish.* **31:** 977–982.

Hirayama, K.
    1966a   Studies on water control by filtration through sand bed in a marine aquarium with closed circulating system. IV. Rate of pollution of water by fish, and the possible number and weight of fish kept in an aquarium. *Bull. Jap. Soc. Sci. Fish.* **32:** 20–26.

Hirayama, K.
    1966b   Influence of nitrate accumulated in culturing water on *Octopus vulgaris. Bull. Jap. Soc. Sci. Fish.* **32:** 105–111.

Hirayama, K.
    1970   Studies of water by control by filtration through sand bed in a marine aquarium with closed circulating system. VI. Acidification of aquarium water. *Bull. Jap. Soc. Sci. Fish.* **36:** 26–34.

Hoar, W. S.
    1975   General and comparative physiology, 2nd ed. Prentice-Hall, Englewood Cliffs, N.J., 848 pp.

Hoff, J. G. and J. R. Westman.
    1966   The temperature tolerances of three species of marine fishes. *J. Mar. Res.* **24:** 131–140.

Hoffman, G. L.
    1967   Parasites of North American freshwater fishes. Univ. Cal. Press, Berkeley, 486 pp.

Hoffman, G. L.
    1974   Disinfection of contaminated water by ultraviolet irradiation, with emphasis on whirling disease (*Myxosoma cerebralis*) and its effect on fish. *Trans. Am. Fish. Soc.* **103:** 541–550.

Hoffman, G. L.
    1975   Whirling disease (*Myxosoma cerebralis*): control with ultraviolet irradiation and effect on fish. *J. Wildl. Dis.* **11:** 505–507.

Hoffman, G. L. and F. P. Meyer.
    1974   Parasites of freshwater fishes. TFH Publ., Neptune City, N.J., 224 pp.

Hoigné, J. and H. Bader.
   1976   The role of hydroxyl radical reactions in ozonation processes in aqueous solutions. *Water Res.* **10:** 377–386.

Honn, K. V. and W. Chavin.
   1976   Utility of ozone treatment in the maintenance of water quality in a closed marine system. *Mar. Biol.* **34:** 201–209.

Honn, K. V., G. M. Glezman, and W. Chavin.
   1976   A high capacity ozone generator for use in aquaculture and water processing. *Mar. Biol.* **34:** 211–216.

Huibers, D. T. A., R. McNabney, and A. Halfon.
   1969   Ozone treatment of secondary effluents from waste-water treatment plants. Report No. TWRC-4, Fed. Water Pollut. Control Adm., Cincinnati, 62 pp.

Jackim, E., J. M. Hamlin, and S. Sonis.
   1970   Effects of metal poisoning on five liver enzymes in the killifish (*Fundulus heteroclitus*). *J. Fish. Res. Board Can.* **27:** 383–390.

Jaffe, E. R.
   1964   Metabolic processes involved in the formation and reduction of methemoglobin in human erythrocytes. *In* The red blood cell, C. Biship and D. Surgenor (eds.). Academic Press, New York, pp. 397–422.

Johnson, P. T.
   1968   An annotated bibliography of pathology in invertebrates other than insects. Burgess, Minneapolis, 322 pp.

Johnson, P. W. and J. McN. Sieburth.
   1974   Ammonia removal by selective ion exchange: a backup system for microbiological filters in closed-system aquaculture. *Aquaculture* **4:** 61–68.

Johnson, S. K.
   1978   Handbook of shrimp diseases, 2nd ed. TAMU-SG-75-603, Texas A & M Univ., College Station, 23 pp.

Jones, J. R. E.
   1938   The relative toxicity of salts of lead, zinc, and copper to the stickleback (*Gasterosteus aculeatus* L.) and the effect of calcium on the toxicity of lead and zinc salts. *J. Exp. Biol.* **15:** 394–407.

Jones, J. R. E.
   1939   The relation between the electrolytic solution pressures of the metals and their toxicity to the stickleback (*Gasterosteus aculeatus* L.). *J. Exp. Biol.* **16:** 425–437.

Jorgensen, S. E., O. Libor, K. L. Graber, and K. Barkacs.
   1976   Ammonia removal by use of clinoptilolite. *Water Res.* **10:** 213–224.

Joyce, R. S. and V. A. Sukenik.
   1964   Feasibility of granular activated-carbon adsorption for waste-water renovation. PHS Publ. No. 999-WP-12, Pub. Health Serv., Cincinnati, 32 pp.

Kabata, Z.
   1970   Diseases of fishes, book 1. Crustacea as enemies of fishes. TFH Publ., Neptune City, N.J., 169 pp.

Kanungo, M. S. and C. L. Prosser.
   1959   Physiological and biochemical adaptation of goldfish to cold and warm temperature. I. Standard and active oxygen consumptions of cold- and warm-acclimated goldfish at various temperatures. *J. Cell. Physiol.* **54:** 259–264.

Kawai, A., Y. Yoshida, and M. Kinata.
  1964   Biochemical studies on the bacteria in aquarium with circulating system. I. Changes of the qualities of breeding water and bacterial population of the aquarium during fish cultivation. *Bull. Jap. Soc. Sci. Fish.* **30:** 55–62.

Kawai, A., Y. Yoshida, and M. Kinata.
  1965   Biochemical studies on the bacteria in aquarium with circulating system. II. Nitrifying activity of the filter sand. *Bull. Jap. Soc. Sci. Fish.* **31:** 65–71.

Kerstetter, T. H. and M. Keeler.
  1976   On the interaction of NH_4^+ and Na^+ fluxes in the isolated trout gill. *J. Exp. Biol.* **64:** 517–527.

Kingsford, E.
  1975   Treatment of exotic marine fish diseases. Pet Ref. Ser. No. 1, Palmetto, St. Petersburg, Fla., 92 pp.

Klontz, G. W.
  1973   Syllabus of fish health management. TAMU-SG-74-401, Texas A & M Univ., College Station, 165 pp.

Knepp, G. W. and G. F. Arkin.
  1973   Ammonia toxicity levels and nitrate tolerance of channel catfish. *Prog. Fish-Cult.* **35:** 221–224.

Konikoff, M. A.
  1974   Comparison of clinoptilolite and biofilters for nitrogen removal in recirculating fish culture systems. *Diss. Abstr. (Zool.)* **34:** 4755–B.

Koon, J. H. and W. J. Kaufman.
  1971   The optimization of ammonia removal by ion exchange using clinoptilolite. SERL Report. No. 71-5, Univ. Cal., Berkeley, 189 pp.

Koon, J. H. and W. J. Kaufman.
  1975   Ammonia removal from municipal wastewaters by ion exchange. *J. Water Pollut. Control Fed.* **47:** 448–465.

Korngold, E.
  1972   Removal of nitrates from potable water by ion exchange. *Water, Air, Soil Pollut.* **2:** 15–22.

Kuhn, P. A.
  1956   Removal of ammonia from sewage effluent. M. S. thesis, Dept. Civ. Engr., Univ. Wis., Madison.

Kunin, R.
  1963   Helpful hints in ion exchange technology. Amber-Hi-Lites, Rohm and Haas Co., Philadelphia, 12 pp.

Kuhl, H. and H. Mann.
  1962   Modellversuche zum Stoffhaushalt in Aquarien bie verschiedenem Salzgehalt. *Kiel Meeresforsch.* **18:** 89–92.

Kuwatani, Y., T. Nishii, and F. Isogai.
  1969   Effects of nitrate in culture water on the growth of the Japanese pearl oyster. *Bull. Natl. Pearl Res. Lab.* **14:** 1735–1747. (Japanese)

Larmoyeux, J. D. and R. G. Piper.
  1973   Effects of water reuse on rainbow trout in hatcheries. *Prog. Fish-Cult.* **35:** 2–8.

Lees, H.
  1952   The biochemistry of the nitrifying organisms. 1. The ammonia-oxidizing systems of *Nitrosomonas. Biochem. J.* **52:** 134–139.

Lenfant, C. and K. Johansen.
  1966 Respiratory function in the elasmobranch *Squalus suckleyi* G. *Respir. Physiol.* **1:** 13–29.

Levine, G. and T. L. Meade
  1976. The effects of disease treatment on nitrification in closed system aquaculture. *Proc. 7th Ann. Meet. World Maricult Soc.*, J. W. Avault Jr. (ed.). World Maricult. Soc., La. State Univ., Baton Rouge, pp. 483–493.

Lewis, S. D. and W. M. Lewis.
  1971 The effect of zinc and copper on the osmolality of blood serum of the channel catfish, *Ictalurus punctatus* Rafinesque, and golden shiner, *Notemigonus crysoleucas* Mitchill. *Trans. Am. Fish. Soc.* **100:** 639–643.

Lippmann, F.
  1960 Versuche zur Aufklarung der Bildungsbedingungen von Calcit und Aragonit. *Fortschr. Mineral.* **38:** 156–161.

Lloyd, R.
  1961 The toxicity of ammonia to rainbow trout (*Salmo gairdneri*). *Waste Water Treat. J.* **8:** 278–279.

Lloyd, R. and L. D. Orr.
  1969 The diuretic response by rainbow trout to sub-lethal concentrations of ammonia. *Water Res.* **3:** 335–344.

Lom, J.
  1969 Cold-blooded vertebrate immunity to protozoa. *In* Immunity to parasitic animals, Vol. 1, G. J. Jackson, R. Herman, and I. Singer (eds.). Appleton-Century Crofts, New York, pp. 249–265.

Maetz, J., P. Payan, and G. de Renzis.
  1976 Controversial aspects of ionic uptake in freshwater animals. *In* Perspectives in experimental biology, Vol. 1, P. S. Davis (ed.). Pergamon, Oxford, pp. 77–92.

Manabe, T.
  1969 New modification of Lubrochinsky's indophenol method for direct microanalysis of ammonia-N in sea water. *Bull. Jap. Soc. Sci. Fish.* **35:** 897–906. (Japanese)

Mangum, C. P., J. A. Dykens, R. P. Henry, and G. Polites.
  1978 The excretion of NH_4^+ and its ouabain sensitivity in aquatic annelids and molluscs. *J. Exp. Zool.* **203:** 151–157.

Maqsood, R. and A. Benedek.
  1977 Low-temperature organic removal and denitrification in activated carbon columns. *J. Water Pollut. Control. Fed.* **49:** 2107–2117.

Martin, D. M.
  1972 Marine chemistry, Vol. 1. Marcel-Dekker, New York, 389 pp.

Martinez, W. W.
  1962 Phosphate and nitrate removal from treated sewage by exchange resins. M. S. thesis, Dept. Civ. Engr., Penn. State Univ., College Park.

Matthiessen, P. and A. E. Brafield.
  1973 The effects of dissolved zinc on the gills of the stickleback *Gasterosteus aculeatus* (L.). *J. Fish Biol.* **5:** 607–613.

McCarthy, J. J. and C. H. Smith.
  1974 A review of ozone and its application to domestic wastewater treatment. *J. Am. Water Works Assoc.* **66:** 718–725.

McCreary, J. J. and V. L. Snoeyink.
    1977   Granular activated carbon in water treatment. *J. Am. Water Works Assoc.* **69:** 437–444.

McKim, J. M., G. M. Christensen, and E. P. Hunt.
    1970   Changes in the blood of brook trout (*Salvelinus fontinalis*) after short-term and long-term exposure to copper. *J. Fish. Res. Board Can.* **27:** 1883–1889.

Merkens, J. C. and K. M. Downing.
    1957   The effect of tension of dissolved oxygen on the toxicity of un-ionized ammonia to several species of fish. *Ann. Appl. Biol.* **45:** 521–527.

Meyers, P. A. and J. G. Quinn.
    1971   Interaction between fatty acids and calcite in seawater. *Limnol. Oceanogr.* **16:** 992–997.

Milne, M. D., B. H. Scribner, and M. A. Crawford.
    1958   Non-ionic diffusion and the excretion of weak acids and bases. *Am. J. Med.* **24:** 709–729.

Morris, J. C. and W. J. Weber Jr.
    1964   Adsorption of biochemically resistant materials from solution. 1. PHS Publ. No. 999-SP-11, 74 pp.

Morris, R. W.
    1962   Body size and temperature sensitivity in the cichlid fish, *Aequidens portalegrensis* (Hensel). *Am. Nat.* **96:** 35–50.

Mullin, J. D. and J. P. Riley.
    1955   The spectrophotometric determination of nitrate in natural waters, with particular reference to sea-water. *Anal. Chim. Acta* **12:** 464–480.

Murphy, J. and J. P. Riley.
    1962   A modified single solution method for the determination of phosphate in natural waters. *Anal. Chim. Acta* **27:** 31–36.

Murray, C. N. and J. P. Riley.
    1969   The solubility of gases in distilled water and in sea water. II. Oxygen. *Deep-Sea Res.* **16:** 311–320.

National Academy of Science.
    1973   Aquatic animal health. Nat. Acad. Sci., Washington, D.C., 46 pp.

Nebel, C., R. D. Gottschling, R. L. Hutchinson, T. J. McBride, D. M. Taylor, J. L. Pavoni, M. E. Tittlebaum, H. E. Spencer, and M. Fleischman.
    1973   Ozone disinfection of industrial-municipal secondary effluents. *J. Water Pollut. Control Fed.* **45:** 2493–2507.

Német, A. G.
    1961   Flow of gas-liquid mixtures in vertical tubes. *Ind. Engr. Chem.* **53:** 151–154.

Nesselson, R. J.
    1954   Removal of inorganic nitrogen from sewage effluent. Ph. D. thesis, Dept. Civ. Engr., Univ. Wis., Madison.

Ng, K. S. and J. C. Mueller.
    1975   Foam separation—a technique for water pollution abatement. *Water Sew. Works.* **122(6):** 48–55.

Nigrelli, R. F.
    1936   Life-history of *Oodinium ocellatum. Zoologica (N.Y.)* **21:** 129–164, 9 pl.

Nigrelli, R. F. and G. D. Ruggieri.
   1966 Enzootics in the New York Aquarium caused by *Cryptocaryon irritans* Brown, 1951 (= *Ichthyophthirius marinus* Sikama, 1961), a histophagous ciliate in the skin, eyes and gills of marine fishes. *Zoologica (N.Y.)* **51:** 97–102, 7 pl.

O'Donovan, D. C.
   1965 Treatment with ozone. *J. Am. Water Works Assoc.* **57:** 1167–1194.

Olson, K. R. and P. O. Fromm.
   1971 Excretion of urea by two teleosts exposed to different concentrations of ambient ammonia. *Comp. Biochem. Physiol.* **40A:** 999–1007.

O'Melia, C. R. and W. Stumm.
   1967 Theory of water filtration. *J. Am. Water Works Assoc.* **59:** 1393–1412.

O'Kelley, J. C.
   1974 Inorganic nutrients. *In* Algal physiology and biochemistry, W. D. P. Steward (ed.). Univ. Cal. Press, Berkeley, pp. 610–635.

Painter, H. A.
   1970 A review of the literature on inorganic nitrogen metabolism in microorganisms. *Water Res.* **4:** 393–450.

Parkhurst, J. D., F. D. Dryden, G. N. McDermott, and J. English.
   1967 Pomona activated carbon pilot plant. *J. Water Pollut. Control Fed.* **39:** R70–R81.

Pauley, G. B. and M. R. Tripp (eds.).
   1975 Diseases of crustaceans. *Mar. Fish. Rev.* **37(5/6):** 1–64.

Payan, P. and J. Maetz.
   1973 Branchial sodium transport mechanisms in *Scyliorhinus canicula:* evidence for Na^+/NH_4^+ Na^+/H^+ exchanges and for a role for carbonic anhydrase. *J. Exp. Biol.* **58:** 487–502.

Payan, P. and A. J. Matty.
   1975 The characteristics of ammonia excretion by a perfused isolated head of trout (*Salmo gairdneri*): effect of temperature and CO_2-free Ringer. *J. Comp. Physiol.* **96:** 167–184.

Perrone, S. J. and T. L. Meade.
   1977 Protective effect of chloride on nitrite toxicity in coho salmon (*Oncorhynchus kisutch*). *J. Fish. Res. Board Can.* **34:** 486–492.

Plummer, L. N. and F. T. Mackenzie.
   1974 Predicting mineral solubility from rate data: application to the dissolution of magnesian calcites. *Am. J. Sci.* **274:** 61–83.

Quastel, J. H. and P. G. Scholefield.
   1951 Biochemistry of nitrification in soil. *Bacteriol. Rev.* **15:** 1–53.

Reichenbach-Klinke, von H-H.
   1967 Untersuchungen über die Einwirkung des Ammoniakgehalts auf den Fischorganismus. *Arch. Fisch.* **17:** 122–132.

Reichenback-Klinke, von H-H. and E. Elkan.
   1965 The principal diseases of lower vertebrates. Academic Press, London, 600 pp.

Ribelin, W. R. and G. Migaki (eds.).
   1975 The pathology of fishes. Univ. Wis. Press, Madison, 1004 pp.

Riley, J. P., D. E. Robertson, J. W. R. Dutton, N. T. Mitchell, and P. J. LeB. Williams.

1975   Analytical chemistry of sea water. *In* Chemical oceanography, Vol. 3, 2nd ed., J. P. Riley and G. Skirrow (eds.). Academic Press, London, pp. 193–514.

Robinette, H. R.
1976   Effect of selected sublethal levels of ammonia on the growth of channel catfish (*Ictalurus punctatus*). *Prog. Fish-Cult.* **38:** 26–29.

Root, R.
1931   The respiratory function of the blood of marine fishes. *Biol. Bull.* **61:** 427–456.

Rosen, H. M.
1973   Use of ozone and oxygen in advanced wastewater treatment. *J. Water Pollut. Control Fed.* **45:** 2521–2536.

Rubin, A. J. and G. A. Elmaraghy.
1977   Studies on the toxicity of ammonia, nitrate and their mixtures to guppy fry. *Water Res.* **11:** 927–935.

Rubin, E., R. Everett Jr., J. J. Weinstock, and H. M. Schoen.
1963   Contaminant removal from sewage plant effluents by foaming. PHS Publ. No. 999-WF-5, Pub. Health Serv., Cincinnati, 56 pp.

Russo, R. C., C. E. Smith, and R. V. Thurston.
1974   Acute toxicity of nitrite to rainbow trout (*Salmo gairdneri*). *J. Fish. Res. Board Can.* **31:** 1653–1655.

Saeki, A.
1958   Studies on fish culture in filtered closed-circulation aquaria. I. Fundamental theory and system design standards. *Bull. Jap. Soc. Sci. Fish.* **23:** 684–695. (*Trans.* E. R. Hope, Dir. Sci. Inf. Serv., Def. Res. Board Can., issued Jan. 1964.)

Saffran, W. A. and Q. H. Gibson.
1976   Kinetics of the Bohr effect of menhaden hemoglobin, *Brevoortia tyrannus*. *Biochem. Biophys. Res. Commun.* **69:** 383–388.

Sander, E.
1967   Skimmers in the marine aquarium. *Petfish Mon.* **2:** 48–51.

Sanders, J. E., J. L. Fryer, D. A. Leith, and K. D. Moore.
1972   Control of the infectious protozoan *Ceratomyxa shasta* by treating hatchery water supplies. *Prog. Fish-Cult.* **34:** 13–17.

Sarig, S.
1971   Diseases of fishes, book 3. The prevention and treatment of diseases of warmwater fishes under subtropical conditions, with special emphasis on intensive fish farming. TFH Publ., Neptune City, N.J., 127 pp.

Saunders, R. L.
1962   The irrigation of the gills of fishes. II. Efficiency of oxygen uptake in relation to respiratory flow activity and concentrations of oxygen and carbon dioxide. *Can. J. Zool.* **40:** 817–862.

Schiøtz, A.
1976   Removal of copper from a saltwater system. *Drum and Croaker* **16:** 44.

Schlesner, H. and G. Rheinheimer.
1974   Auswirkungen einer Ozonisierungsanlage auf den Bakteriengehalt des Wassers eines Schauaquariums. *Kiel. Meeresforsch.* **30:** 117–129.

Schlieper, C.
  1950   Temperaturbezogne Regulationen des Grundumsatzes bei Wechseiwarmen Tierren. *Biol. Zentralbl.* **69:** 216–227.

Schreckenbach, K., R. Spangenberg, and S. Krug.
  1975   Cause of gill necrosis. *Z. Binnefish, DDR* **22:** 257–288 (German)

Segedi, R. and W. E. Kelley.
  1964   A new formula for artificial sea water. *In* Sea-water systems for experimental aquariums: a collection of papers, J. R. Clark and R. L. Clark (eds.). Res. Report 63, U. S. Dept. Inter., Bur. Sport Fish. Wildl., pp. 17–19.

Sellers, C. M. Jr., A. G. Heath, and M. L. Bass.
  1975   The effect of sublethal concentrations of copper and zinc on ventilatory activity, blood oxygen and pH in rainbow trout (*Salmo gairdneri*). *Water Res.* **9:** 401–408.

Shechter, H. and N. Gruener.
  1976   An evaluation of the ion selective electrode method for determination of nitrate in highly mineralized drinking water. *J. Am. Water Works Assoc.* **68:** 543–546.

Sheldon, R. W., T. P. T. Evelyn, and T. R. Parsons.
  1967   On the occurrence and formation of small particles in seawater. *Limnol. Oceanogr.* **12:** 367–375.

Sindermann, C. J.
  1970   Bibliography of diseases and parasites of marine fish and shellfish, with emphasis on commercially important species. Informal Report No. II, U. S. Dept. Inter., Trop. Atlan. Biol. Lab., Gulf and Atl. Reg., Miami, 440 pp.

Sindermann, C. J. (ed.).
  1977   Disease diagnosis and control in North American marine aquaculture. Elsevier, Amsterdam, 330 pp.

Singer, P. C. and W. B. Zilli.
  1975   Ozonation of ammonia in wastewater. *Water Res.* **9:** 127–134.

Skidmore, J. F.
  1964   Toxicity of zinc compounds to aquatic animals, with special reference to fish. *Quart. Rev. Biol.* **39:** 227–247.

Skirrow, G.
  1975   The dissolved gases—carbon dioxide. *In* Chemical Oceanography, Vol. 2, 2nd ed., J. P. Riley and G. Skirrow (eds.). Academic Press, London, pp. 1–192.

Smith, C. E.
  1972   Effects of metabolic products on the quality of rainbow trout. *Am. Fish. U. S. Trout News* **17:** 1–3.

Smith, C. E. and R. G. Piper.
  1975   Lesions associated with chronic exposure to ammonia. *In* The pathology of fishes, W. E. Ribelin and G. Migaki (eds.). Univ. Wis. Press, Madison, pp. 497–514.

Smith, C. E. and W. G. Williams.
  1974   Experimental nitrite toxicity in rainbow trout and chinook salmon. *Trans. Am. Fish. Soc.* **103:** 389–390.

Smith, R. G. Jr. and H. L. Windom.
  1972   Analytical handbook for the determination of arsenic, cadmium, cobalt, cop-

per, iron, lead, manganese, mercury, nickel, silver and zinc in the marine and estuarine environments. Tech. Report Ser. 72–6, Ga. Mar. Sci.' Cen., Univ. Ga., Savannah, 62 pp.

Snieszko, S. F. (ed.).
1970 A symposium on diseases of fishes and shellfishes. Spec. Publ. No. 5, Am. Fish. Soc., Washington, D.C., 526 pp.

Solórzano, L.
1969 Determination of ammonia in natural waters by the phenolhypochlorite method. *Limnol. Oceanogr.* **14:** 799–801.

Somero, G. N. and P. W. Hochachka.
1976 Biochemical adaptations to temperature. *In* Adaptation to environment: essays on the physiology of marine animals, R. C. Newell (ed.). Butterworths, London, pp. 125–190.

Sontheimer, H.
1974 Use of activated carbon in water treatment practice and its regeneration. Spec. Subj. 3, Lehrstuhl für Wasserchimie der Universität Karlsruhe, West Germany, 6 pp.

Sousa, R. J. and T. L. Meade.
1977 The influence of ammonia on the oxygen delivery system of coho salmon hemoglobin. *Comp. Biochem. Physiol.* **58A:** 23–28.

Spotte, S.
1970 Fish and invertebrate culture: water management in closed systems, 1st ed. Wiley, New York, 145 pp.

Spotte, S.
1973 Marine aquarium keeping: the science, animals, and art. Wiley, New York, 173 pp.

Spotte, S.
1974 Aquarium techniques: closed-system marine aquariums. *In* Experimental marine biology, R. N. Mariscal (ed.). Academic Press, New York, pp. 1–19.

Srna, R. F. and A. Baggaley.
1975 Kinetic response of perturbed marine nitrification systems. *J. Water Pollut. Control Fed.* **47:** 472–486.

Srna, R. F., C. Epifanio, M. Hartman, G. Pruder, and A. Stubbs.
1973 The use of ion specific electrodes for chemical monitoring of marine systems. I. The ammonia electrode as a sensitive water quality indicator probe for recirculating mariculture systems. DEL-SG-73, Coll. Mar. Stud., Univ. Del., Newark, 20 pp.

Stainton, M. P., M. J. Capel, and F. A. J. Armstrong.
1977 The chemical analysis of fresh water, 2nd ed. Can. Fish. Mar. Serv. Misc. Spec. Publ. 25, 180 pp.

Strickland, J. D. H. and T. R. Parsons.
1972 A practical handbook of seawater analysis, 2nd ed. Bull. 167, Fish. Res. Board Can., Ottawa, 310 pp.

Suess, E.
1970 Interaction of organic compounds with calcium carbonate. I. Association phenomena and geochemical implications. *Geochim. Cosmochim. Acta* **34:** 157–168.

Sutcliffe, W. H., E. R. Baylor, and D. W. Menzel.
1963 Sea surface chemistry and Langmuir circulation. *Deep-Sea Res.* **10:** 233–243.

Tanaka, T., K. Hiiro, and A. Kawahara.
  1977 Determination of phosphate ion using lead ion-sensitive electrode. Z. Anal. Chem. **286:** 212–213.

Tchobanoglous, G.
  1970 Filtration techniques in tertiary treatment. J. Water Pollut. Control Fed. **42:** 604–623.

Tchobangoglous, G. (ed.).
  1972 Wastewater engineering: collection, treatment, disposal. McGraw-Hill, New York, 782 pp.

Tomlinson, T. G. A., G. Boon, and G. N. A. Trotman.
  1966 Inhibition of nitrification in the activated sludge process of sewage disposal. J. Appl. Bacteriol. **29:** 266–291.

Trussell, R. P.
  1972 The percent un-ionized ammonia in aqueous solutions at different pH levels and temperatures. J. Fish. Res. Board Can. **29:** 1505–1507.

Tyler, A. V.
  1966 Some lethal temperature relations of two minnows of the genus Chrosomus. Can. J. Zool. **44:** 349–361.

Vernon, E. H.
  1954 The toxicity of heavy metals to fish, with special reference to lead, zinc and copper. Can. Fish Cult. **15:** 1–6.

Vlasenko, M. I.
  1969 Ultraviolet rays as a method for the control of diseases of fish eggs and young fishes. J. (Prob.) Ichthyol. **9:** 697–705.

Warren, K. S.
  1962 Ammonia toxicity and pH. Nature **195:** 47–49.

Watson, T. A. and B. A. McKeown.
  1976 The effect of sublethal concentrations of zinc on growth and plasma glucose levels in rainbow trout, Salmo gairdneri Richardson. J. Wildl. Dis. **12:** 263–270.

Weatherley, A. H.
  1970 Effects of superabundant oxygen on thermal tolerance of goldfish. Biol. Bull. **139:** 229–238.

Weber, W. J. and W. Stumm.
  1963 Buffer systems of natural fresh waters. J. Chem. Engr. Data **8:** 464–468.

Wedemeyer, G. A. and N. C. Nelson.
  1977 Survival of two bacterial fish pathogens (Aeromonas salmonicida and the enteric redmouth bacterium) in ozonated, chlorinated, and untreated waters. J. Fish. Res. Board Can. **34:** 429–432.

Wellborn, T. L. Jr. and W. A. Rogers.
  1966 A key to the common parasitic protozoans of North American fishes. Zool.-Entomol. Dept. Ser. Fish. No. 4, Agri. Exp. Sta., Auburn Univ., Auburn, 17 pp.

Wells, A. N.
  1935 The influence of temperature upon the respiratory metabolism of the Pacific killifish, Fundulus parvipinnis. Physiol. Zool. **8:** 196–227.

Westin, D. T.
  1974 Nitrate and nitrite toxicity to salmonid fishes. Prog. Fish-Cult. **36:** 86–89.

Weyl, P. K.
> 1967   The solution behavior of carbonate materials in sea water. Studies Trop. Oceanogr., Miami **5**: 178–228.

Weyl, P. K.
> 1970   Oceanography: an introduction to the marine environment. Wiley, New York, 535 pp.

Wheaton, F. W.
> 1977   Aquacultural engineering. Wiley, New York, 708 pp.

Whitfield, M.
> 1974   The hydrolysis of ammonium ions in sea water—a theoretical study. *J. Mar. Biol. Assoc. U. K.* **54**: 565–580.

Williams, J.
> 1962   Oceanography. Little, Brown, Boston, 242 pp.

Wooster, W. S., A. J. Lee, and G. Dietrich.
> 1969   Redefinition of salinity. *J. Mar. Res.* **27**: 358–360.

Wuhrmann, K. von and H. Woker.
> 1948   Beiträge zur Toxikologie der Fische—II. Experimentelle untersuchungen über die Ammoniak- und Blausäurevergiftung. *Schweiz. Z. Hydrol.* **11**: 210–244.

Wuhrmann, K. von and H. Woker.
> 1953   Beiträge zur Toxikologie der Fische—VIII. Temperatur auf Fische. *Schweiz. Z. Hydrol.* **15**: 235–260.

Yao, K. M., M. T. Habibian, and C. R. O'Melia.
> 1971   Water and waste water filtration: concepts and applications. *Environ. Sci. Tech.* **5**: 1105–1112.

Yoshida, Y.
> 1967   Studies on the marine nitrifying bacteria: with special reference to characteristics and nitrite formation of marine nitrite formers. *Bull. Misaki Mar. Biol. Inst.*, Kyoto Univ., Kyoto, **11**: 1–58.

Zerbe, W. B. and C. B. Taylor.
> 1953   Sea water temperature and density reduction tables. Spec. Publ. 298, U. S. Coast Geodetic Surv., U. S. Dept. Comm., 21 pp.

ZoBell, C. E. and H. D. Michener.
> 1938   A paradox in the adaptation of marine bacteria to hypertonic solutions. *Science* **87**: 328–329.

Index

Absorbance, 129, 131-134
Acclimation temperature, 87, 88
Acetone, 133, 134
Acid-base imbalance, 87, 115
Acidemia, 115
Acidity, of cells, 83
   of water, 104, 105, 110
Activated carbon, activation of, 41
   DOC removal by, 40-43, 110
   and element depletion, 94
   and heavy metal removal, 126
   manufacture of, 40, 41
   see also Granular activated carbon
Activated sludge, heavy metals in 5, 126
Active transport, definition of, 93
   effect of oxygen, 94
Adsorbate, concentration of, 42
   definition of, 40
   mass transfer of, 41, 42
   uptake by GAC, 42
Adsorbent, definition of, 40
   and GAC, 42
Adsorption, definition of, 40
   DOC onto calcite, 108
   see also Adsorption rate; Physical
     adsorption
Adsorption rate, DOC onto GAC, 42, 43
Adsorptive capacity, of GAC, 59
   of ion exchangers, 52-61
Aequidens portalegrensis, see Cichlid
Aeration, and ammonia toxicity, 125
   of artificial seawater, 95, 97
   before ion exchange, 59
   definition of, 74
   during thermal acclimation, 88
   and heavy metal toxicity, 126

   after ozonation, 73
   of raw water, 121
   supplementary, 86
   of trace salt solutions, 102
Aeromonas salmonicida, 69
Aged tap water, definition of, 20
Aggregate, definition of, 21
   formation of, 21, 22
   removal of, 24
Airborne toxicants, 96
Air bubbles, in airlifts, 76, 87
   DOC adsorption by, 21-23, 40, 47, 48
   in foam fractionators, 48-50
   in gas exchange, 74
Air compressor, 49, 50, 77
Air diffusers, in foam fractionators, 50. *See
   also* Airstones
Air injection, into airlifts, 75-78, 87
   into foam fractionators, 49, 50, 59
Airlift, design of, 12, 78-80
   detritus formation in, 22
   with foam fractionators, 50, 59
   with GAC contactors, 43, 44, 46
   and gas exchange, 12
   operating principles of, 75-78
   operation of, 87
   and seawater mixing vats, 97, 100
   turbulence in, 87
Airlift pumps, *see* Airlift
Airline tubing, 97, 136
Airstones, in airlifts, 78, 86, 87
   in trace salt solutions, 102
   see also Air diffusers
Airstripping, *see* Foam fractionation
Airstripping towers, ammonia removal by,
   48

161

Air volume, in airlifts, 76, 77, 80
in foam fractionators, 48
Algae, and clogged filter sleeves, 36
elements essential to, 94, 95
exudates of, 94
trace ion removal by, 59
and turbidity, 21
Alkali-iodide-azide solution, 143
Alkaline solution, 131
Alkalinity, decline in, 108, 110
definition of, 104, 134
determination of, 134, 135
of freshwater, 104
heavy metal precipitation by, 118
and ion exchange, 51-53
maintenance of, 107, 110
and nitrate toxicity, 117
of seawater, 104
Alkalinization, 106
Aluminum foil, 128, 130
Ambient temperature, see Acclimation temperature
American oyster, 117
Amino acids, from mineralization, 3
synthesis of, 93
Ammonia, and alkalinization, 106
as ammonia nitrogen, 129
atmospheric, 131
in blood, 113
in conditioned aquariums, 10, 11
determination of, 127, 129-132
from dissimilation, 10, 11
effect on hemoglobin, 114
effect on immunity, 119, 120
elevation in shipping water, 88
entry into gills, 113, 114
excretion by gills, 113
excretion rate of, 114
and foam fractionation, 48
forced retention of, 114
hydrolysis of, 112, 113
inhibition to Nitrobacter, 11
maximum limit of, 125
from mineralization, 112
in new aquariums, 18
oxidation by ozone, 66
pH effect on oxidation, 6
production of, 3, 7

reduction by water changes, 125
removal by ion exchange, 40, 55
species of, 3
temperature effect on oxidation, 3, 6
tissue penetration of, 113, 114
toxicity of, 3, 112-115, 124
toxicity and aeration, 125
toxicity and anoxia, 114
toxicity and oxygen, 114, 115
toxicity and pH, 114
see also Ammonium ion; Free ammonia
Ammonia oxidizers, 7
Ammonium ion, definition of, 3
exchange for sodium, 113
hydrolysis of, 112-114, 124
oxidation and pH, 108, 109
percentage of, 112, 113, 129
removal by ion exchange, 51-53, 55-57
see also Ammonia; Ammonium ion hydrolysis; and Free ammonia
Ammonium ion hydrolysis, pH effect on, 113, 114
pK of, 112
salinity effect on, 113
temperature effect on, 113
see also Ammonium ion
Ammonium molybdate, 128
Ammonium molybdate solution, 128
Ammonium sulfate, 131
Ammonotelic, definition of, 113
Analytical balance, 103, 127
Analytical methods, 127-144
Animal bone, 40
Animal flesh, 123
Anion exchangers, basic chloride, 57
nitrate removal by, 57
phosphate removal by, 57
substances removed by, 53
types of, 51
Anoxic conditions, and ammonia toxicity, 114
and carrying capacity, 16
and dissimilation, 9
from poor circulation, 19
Anthracite coal, 25
Antibacterial agents, effects on nitrification, 4, 5
Antibiotics, and disinfection, 62